APPLIED MATH
WITH PYTHON

APPLIED MATH
WITH PYTHON

Solve Real-World Problems with
Python-Based Solutions

BLAKE RAYFIELD

WILEY

Library of Congress Control Number: 2026936342

Print ISBN: 9781394370757

ePDF ISBN: 9781394370771

ePub ISBN: 9781394370764

Obook ISBN: 9781394370788

Cover Design: Wiley
Cover Image: © CSA-Printstock/Getty Images

Printed and bound by CPI Group (UK) Ltd, Croydon, CR0 4YY

C9781394370757_170426

to Theodore and Emery

ACKNOWLEDGMENTS

I have been fortunate to work with a team of dedicated professionals to create this technical book. A special thanks goes to my agent, Jess Haberman, for her representation and steadfast support in bringing this project to life.

Jim Minatel, acquisitions editor, and Pete Gaughan, managing editor, provided the essential leadership and vision required to bring this title to market. It was a pleasure to work with Brad Jones, project manager, who managed the complex process (and me) that got us from a rough outline to a finished manuscript. I also want to thank Dhilip Kumar Rajendran, the content refinement specialist, as well as Kezia Endsley for her expertise in copyediting.

I am especially grateful for Matthew Housley's expertise as technical editor. In addition to catching my subtle and not-so-subtle errors, his close reading and technical insights regarding the code and mathematical concepts ensured the accuracy required for a book of this depth.

Finally, I must thank my family. My wife, Lois Rayfield, and our two children, Emery and Theodore, have supported me and provided the encouragement I needed to get this manuscript across the finish line.

—BLAKE RAYFIELD

ABOUT THE AUTHOR

Blake Rayfield is an assistant professor of Finance at the University of North Florida, with a strong background in higher education, research, and data science. He earned both his MS and PhD in Financial Economics from the University of New Orleans. His work has been published in several peer-reviewed journals, including the *Journal of Financial Research*, the *Quarterly Review of Economics and Finance*, and the *Review of Behavioral Finance*. His research interests focus on Corporate Finance and Investments, with particular emphasis on applying advanced data science methods to financial research.

ABOUT THE AUTHOR

ABOUT THE TECHNICAL EDITOR

Matt Housley holds a PhD in mathematics from the University of Utah and coauthored the book *Fundamentals of Data Engineering* (O'Reilly Media). With Tony Baer, he cohosts *It's About Data,* a podcast covering the latest developments in data, artificial intelligence, technology, and business. Matt enjoys training a new generation of data engineers and currently works as a freelance trainer and consultant.

CONTENTS

INTRODUCTION   *XIX*

PART 1: GETTING STARTED

**CHAPTER 1: INTRODUCTION TO PYTHON FOR BUSINESS
APPLICATIONS**   3

Introducing Python for Business   3
Why Python, Not a Spreadsheet?   4
Setting Up Your Tools   5
    Install Python with the Anaconda Distribution
    (Running Python on Your Machine)   5
    Launch Jupyter Notebook   6
    Cloud-friendly Alternatives   6
The Python Ecosystem   7
    What Is a (Jupyter) Notebook?   8
    Installing Libraries Locally or in a Notebook   8
Writing Your First Python Script   9
Summary   10
Continue Your Learning   10

CHAPTER 2: BASIC MATHEMATICAL OPERATIONS IN PYTHON   11

Numbers, Variables, and Functions: The Foundations
of Business Logic   11
    Understanding Variables   12
    Arithmetic in Python   13
    Working with the *math* Module   14
Data Types in Python   14
    Core Data Types   14
    Why Data Types Matter   16
    Converting Between Types   17
Business Data Structures: Arrays and Matrices   18
    One-dimensional Arrays   18
    Matrices: Two-dimensional Arrays   19
Data Manipulation Basics with Pandas   23
    Constructing a DataFrame   23
    First Looks: *head(), info(), describe()*   24

Working with Columns and Rows	24
Filtering with Booleans	25
Creating New Columns	25
Grouping and Aggregation	26
Joins and Merges	27
Reshaping: Pivot, Melt, Stack	28
Summary	**28**
Continue Your Learning	**28**

CHAPTER 3: VISUALIZATION FOR BUSINESS DECISION-MAKING — 29

The Landscape of Visualization Tools in Python	**29**
Visualization Applications: Dashboarding Frameworks	30
Choosing the Right Visualization Tool for Your Work	31
Graphing Basics with Matplotlib	**32**
Understanding the Structure of a Plot	32
Creating and Working with Plots	33
Customizing Visualizations to Enhance Understanding	35
Plotting Options	36
Creating Effective Visuals to Communicate Business Data	**37**
Time-series Data and Line Charts	38
Cross-sectional Data and Bar or Pie Charts	38
Relational Data and Scatterplots	39
Other Charts You Can Create	41
Visualizing Trends and Patterns for Business Insights	**42**
Highlighting Seasonality and Long-term Growth	42
Comparing Categories and Segments	44
Visualizing Cumulative Effects	46
Smoothing Trends with Rolling Averages	47
Line Charts with Confidence Intervals Using Seaborn	49
Analyzing Relationships and Distributions with *jointplot*	52
Summary	**55**
Continue Your Learning	**55**

PART 2: APPLYING THE MATH

CHAPTER 4: LINEAR ALGEBRA FOR BUSINESS AND FINANCE — 59

Working with Vectors and Matrices	**59**
Understanding Vectors	60
Understanding Matrix	61

Operations with Vectors and Matrices 62
 Scalar Multiplication 63
 The Dot Product 63
 Norms (Vector Lengths) 64
 Combining Matrices 64
 Slicing Matrices 65
 Matrix Multiplication 66
 Transpose 67
Creating and Manipulating Vectors (and Matrices) with NumPy 67
 Step 1: Compute Asset Returns from Prices 69
 Step 2: Portfolio with Constant Weights 70
 Step 3: Portfolio with Time-varying Weights 72
 Comparing Strategies (Same Math, Different Inputs) 75
Eigenvalues and Eigenvectors: Business Applications 76
 What Eigenvalues and Eigenvectors Represent 76
 Why Eigenvalues Matter for Long-term Stability 77
Summary 80
Continue Your Learning 80

CHAPTER 5: CALCULUS FOR BUSINESS PROBLEM SOLVING 83

Numerical Differentiation and Integration in Business Analytics 84
 The Derivative: Finding the Rate of Change 84
 The Second Derivative: Pinpointing the Point of Diminishing Returns 86
 The Integral: Accumulating the Totals 87
The Calculus Ecosystem in Python 90
 Numerical Calculus with NumPy 90
 Symbolic Calculus with SymPy 91
 Advanced Numerical Methods with SciPy 92
 Choosing the Right Tool 93
Solving Business Growth and Pricing Models with Differential
Equations 93
Sensitivity Analysis with Partial Derivatives 96
Case Study: Revenue, Cost, and Profit Analysis 98
 Step 1: Understanding Marginal Cost (the Derivative of Cost) 99
 Step 2: Understanding Marginal Revenue (the Derivative of Revenue) 100
 Step 3: Finding the Sweet Spot with Marginal Profit 102
Summary 104
Continue Your Learning 104

CHAPTER 6: OPTIMIZATION TECHNIQUES FOR BUSINESS STRATEGY 107

The Python Optimization Ecosystem 108
A Framework for Solving Most Optimization Problems 109
The Four-step Formulation Process 109
Understanding the Local vs. Global Optima Issue 110
Applying the Framework: Profit Maximization 110
Linear Programming 112
Constrained Optimization 116
The Geometry of Optimization 116
Visualizing the Difference Between Constrained and Unconstrained Optimization 119
Real-world Applications 122
Portfolio Allocation 122
Supply Chain and Operations 128
Integer Programming for Workforce Scheduling 131
Summary 134
Continue Your Learning 134

CHAPTER 7: PROBABILITY AND STATISTICS FOR BUSINESS ANALYTICS 137

The Python Statistics Ecosystems 137
Understanding Random Variables and Distributions in Business Contexts 138
Discrete vs. Continuous Distributions 139
The Most Common Business Distributions 140
Hypothesis Testing 144
Test Statistics 145
The p-value 146
The A/B Test 147
Confidence Intervals: The Other Side of the Coin 148
Linear Regression 149
Analyzing Marketing Effectiveness 151
Explaining Financial Risk Factors 153
Other Considerations 155
Logistic Regression 156
Predicting Customer Churn 156
Forecasting 161
Summary 164
Continue Your Learning 164

CHAPTER 8: APPLIED BUSINESS PROBLEMS WITH MATH AND
PYTHON   167

Building a Dynamic Loan Amortization Engine   168
Building a Simple Recommender System   171
Maximizing Yield with Constrained Optimization   173
Quality Control with Hypothesis Testing   177
Predicting Employee Attrition with Logistic Regression   179
Summary   185
Continue Your Learning   185

PART 3: VISUALIZING THE NUMBERS

CHAPTER 9: ILLUSTRATING TIME-SERIES AND LINEAR DATA   189

Understanding Your Data Structure   189
    Cross-sectional Data   190
    Time-series Data   192
    Panel Data   193
Visualizing Change Over Time (Time-series)   194
    Time-series Diagnostics   195
    Seasonality and Autocorrelation   201
Panel Data   206
Summary   208
Continue Your Learning   209

CHAPTER 10: ILLUSTRATING CROSS-SECTIONAL DATA   211

Data Categories   211
    The Pie Chart   211
    Donut Charts   213
    Stacked Bar Charts   216
Correlations and Distributions   217
    Bar Charts   218
    Boxplots   220
Correlations in the Cross Section   222
    Scatterplots   222
    Correlation Heatmaps   225
    The Pair Plot   227
Summary   229
Continue Your Learning   230
    Essential Cross-sectional Functions   230

CHAPTER 11: ILLUSTRATING ALTERNATIVE DATA TYPES 233

Textual Analysis 233
The Word Cloud 234
N-grams 236
Visualizing Customer Sentiment 239
Geospatial Data 242
The Choropleth Map 243
The Marker Map 244
The Heatmap 246
Visualizing Networks 248
Visualizing Structure 249
Weighted Graphs 252
Summary 254
Continue Your Learning 254

INDEX 257

INTRODUCTION

In today's data-driven economy, the ability to apply mathematical concepts to solve real-world business problems has become essential across industries. From finance and marketing to operations and strategic planning, organizations rely on data-driven insights to stay competitive. Python, with its powerful libraries and widespread adoption, has emerged as the go-to language for addressing these applied mathematics challenges. Readers will explore and apply a wide range of math topics, from optimization, probability, and statistics to linear algebra and calculus. Each chapter is structured around practical use with several examples, such as optimizing supply chains, forecasting financial performance, or uncovering insights from customer data, making it an essential resource for anyone looking to harness math for success.

WHAT DOES THIS BOOK COVER?

This book is designed to bridge the gap between abstract mathematical theory and practical business results using Python. By combining hands-on code with data-driven charts, you can move beyond simple programming to show you how to use math to solve business challenges. The following sections provide a roadmap of our journey, beginning with a streamlined setup and ending with advanced visualization techniques that drive decision-making.

> **Part 1: Getting Started**   Part 1 hits the ground running, designed to get you up to speed. It provides a streamlined setup of the essential Python toolkit and a concise refresher on the core operations needed to analyze business data immediately. You learn some general programming tricks as well as the libraries and initial visualizations that drive early business insights.
>
> In Chapter 1, "Introduction to Python for Business Applications," you are introduced to the core Python commands and why the language is essential for modern business analytics and automation. Chapter 2, "Basic Mathematical Operations in Python," builds on this by reviewing essential math routines, variables, and data types using powerful libraries like NumPy and Pandas. Chapter 3, "Visualization for Business Decision-making," explores how to apply these foundations to present data in a relatable way that drives organizational decisions.
>
> **Part 2: Applying the Math**   In Part 2, the focus is on the specific mathematical concepts that solve real-world problems. It bridges the gap between theory and application, teaching you how to use Python to optimize operations, model financial risks, and forecast performance. By understanding the underlying math, you will be better equipped to apply calculus, linear algebra, and statistics to achieve concrete business goals.

Chapter 4, "Linear Algebra for Business and Finance," uses vectors and matrices to create a practical language for organizing information and modeling complex financial risks. Chapter 5, "Calculus for Business Problem Solving," provides a toolkit for understanding operational momentum, such as identifying the exact point of diminishing returns in a marketing campaign. Chapter 6, "Optimization Techniques for Business Strategy," and Chapter 7, "Probability and Statistics for Business Analytics," introduce the necessary tools to maximize operational efficiency and manage the uncertainties inherent in business forecasting. This part concludes with Chapter 8, "Applied Business Problems with Math and Python," which synthesizes these tools through real-world case studies in logistics, finance, and operations management.

Part 3: Visualizing the Numbers   Data has more value when it drives decisions. Part 3 focuses on translating complex analysis into clear, compelling visuals that stakeholders can act on. It moves beyond basic charting to specific business use cases, tracking financial trends over time, comparing cross-sectional market performance, and handling alternative data types, ensuring your analysis tells a story that leads to actionable strategy.

Chapter 9, "Illustrating Time-series and Linear Data," focuses on matching the right chart to your data structure and smoothing temporal data to reveal hidden trends. Chapter 10, "Illustrating Cross-sectional Data," teaches you to analyze rank, distribution, and correlation at a single point in time through visual "snapshots" like histograms and scatterplots. Finally, Chapter 11, "Illustrating Alternative Data Types," explores how to visualize unstructured sources like text, geographic maps, and complex networks by converting qualitative signals into visualizations.

WHO SHOULD READ THIS BOOK

This book is for business professionals exploring AI, data science, finance, and business analytics who want a practical, hands-on approach to applying math through Python. It's designed for entrepreneurs, analysts, and managers who may have studied math in the past but need a clear, concise refresher to solve real-world business problems. For busy professionals with limited time, this book offers a streamlined approach, focusing on "just enough math" to drive data-driven decisions and achieve business goals without returning to the classroom.

Instead of overwhelming you with theory and proofs, you'll find actionable examples and Python applications tailored for business leaders, financial analysts, and data-driven decision-makers. The information presented in this book bridges the gap between understanding mathematical concepts and applying them effectively in real-world business scenarios. Even though GenAI has greatly enhanced how businesses analyze data, you still need to bridge the knowledge gap to make the most of GenAI's capabilities. You will learn to optimize operations, forecast financial performance, and uncover other insights from complex data.

READER SUPPORT FOR THIS BOOK

Companion Download Files

The book mentions some additional files, such as CSV files or spreadsheets. These items are available for digital download from `http://www.wiley.com/go/AppliedMathWithPython` or at `https://github.com/bkrayfield/Applied-Math-With-Python`.

How to Contact the Author

I appreciate your input and questions about this book! Email me at `blake.rayfield@outlook.com`, or message me on LinkedIn at `https://www.linkedin.com/in/blake-rayfield/`.

PART 1
Getting Started

➤**Chapter 1:** Introduction to Python for Business Applications

➤**Chapter 2:** Basic Mathematical Operations in Python

➤**Chapter 3:** Visualization for Business Decision-making

1

Introduction to Python for Business Applications

This chapter introduces the role of Python in modern business contexts. By the end, you'll understand why Python has become one of the most essential tools in business analytics, decision-making, and automation.

INTRODUCING PYTHON FOR BUSINESS

In today's fast-paced business world, organizations face an increasing volume of data and shrinking timelines for decision-making. Business leaders are under constant pressure to move faster, forecast better, and justify decisions with hard evidence. Whether you're in finance, marketing, operations, or product strategy, data-driven thinking has become non-negotiable. But turning data into insight requires more than dashboards; it requires math.

Python's rise to prominence is no accident. It's open-source (meaning everyone can see how it works), readable, and supported by a robust community of developers. What started as a language for scripting has evolved into a comprehensive ecosystem used across industries such as finance, logistics, retail, tech, and healthcare.

Unlike traditional business tools that are rigid or require expensive customization, Python is agile. With just a few lines of code, business users can:

- ➤ Automate tedious tasks like generating reports.
- ➤ Clean, transform, and analyze datasets.
- ➤ Visualize trends, patterns, and outliers.
- ➤ Apply predictive analytics and machine learning.

To see the difference in practice, picture a marketing analyst who wants to understand how different campaigns impacted sales across regions. In Excel, this might mean multiple sheets, pivot tables, and manual steps. In Python, they can load the data, group it by region and campaign, visualize the trends, and even run a regression analysis to estimate campaign return on investment (ROI) in under 30 lines of code.

The same 30-line script can be scheduled to run every night, updated automatically when new data lands, and even emailed to stakeholders while you sleep. That's "doing more" without adding hours to the day. And because the script is plain text, you can share it and improve it over time without breaking a single cell reference.

This book isn't about theory. You won't be asked to derive formulas or memorize rules. Instead, you'll learn how to apply math through practical, hands-on examples using Python.

You'll walk through business scenarios that look like these:

- ➤ You're a marketing manager trying to segment your customers.
- ➤ You're a finance analyst modeling revenue growth.
- ➤ You're a product lead deciding how to allocate limited resources for maximum impact.

Each chapter covers a math concept (like probability, optimization, or linear algebra), walks through a real-world business use case, and shows how to implement it in Python. Just what you need to make smarter decisions, faster.

WHY PYTHON, NOT A SPREADSHEET?

Spreadsheets are everywhere in business, for good reason. They're fast, familiar, and flexible. But as the questions get more complex, spreadsheets start to show their limitations:

- ➤ Have you ever tried to run a scenario analysis with 20 variables in Excel?
- ➤ Have you ever worked with a dataset that has millions of rows?
- ➤ Have you ever needed to visualize customer behavior in five dimensions?

Python scales where spreadsheets break. With just a few lines of Python code, you can simulate business scenarios, automate repetitive processes, or run robust statistical tests. You can pull data from APIs, merge it with internal tables, train a predictive model, and push the result straight into a dashboard, without ever opening a single workbook.

And here's the best part: Python is modular. You don't have to build from scratch. Need natural language processing? Import *spaCy*. Time-series forecasting? *prophet* or *statsmodels* has you covered. Optimization? *scipy.optimize* or *cvxpy* is ready to go. This book covers these tools and more as you progress with learning. The ecosystem is your toolkit, and most of it is free.

Think of it like this: Excel is great when you're dealing with tables and formulas. But what if you need to process 10,000 customer reviews? Or scrape real-time pricing data from competitor websites? Or train a model to predict which clients are likely to churn? That's where Python shines.

In the chapters to come, you see Python tackle problems that would bring even the most carefully crafted spreadsheet to its knees. You learn how to move from a proof-of-concept notebook to a reliable script, and you discover that a little code goes a long way.

SETTING UP YOUR TOOLS

Before we dive into applying math with Python, let's make sure your environment is ready to go. The good news is that modern Python tools make this process relatively painless. Once you are set up, you will have everything you need to follow along with the examples and exercises in this book.

There are several ways to work with Python. The "best" route depends on your comfort with installing software or if you prefer running code on your own machine or in the cloud. The following sections outline a step-by-step path that works for almost everyone, followed by a few alternatives if you want something lighter or more portable.

Install Python with the Anaconda Distribution (Running Python on Your Machine)

Anaconda is generally found to be the most straightforward path to installing Python on a machine. Anaconda is a free, open-source Python distribution that bundles the language with nearly every library you'll use—Pandas, NumPy, Matplotlib, SciPy, scikit-learn, Jupyter Notebook, and dozens more. Think of it as the "batteries-included" version of Python for applied math.

Why Anaconda makes life easier:

➤ It ships with hundreds of preinstalled packages used in data science and business analytics, so you're not hunting and pecking with the Pip Installs Packages tool (pip) on day one.

➤ It includes Jupyter Notebook out of the box, giving you an interactive, browser-based coding workspace that's perfect for step-by-step math exploration.

➤ It sidesteps most dependency headaches. Anaconda's package manager (conda) keeps library versions compatible.

➤ It works the same on Windows, macOS, and Linux, which means fewer "but it runs on my machine" surprises.

To install Anaconda, do the following:

1. Head to anaconda.com and download the latest free edition.

2. Download the free installer for your operating system. Pick the Python 3.x version (whatever is labeled "Latest"). By choosing "Distribution Installer" you will be not only installing Python, but also an ecosystem of commonly used packages (Recommend). If you want to install a smaller version of Python with fewer preinstalled packages, choose Miniconda. You can install additional packages later as you need them.

3. Run the installer.

4. When the installer finishes, open Anaconda Navigator if you installed the full distribution (or the terminal if you installed Miniconda). The Navigator is a friendly hub that lets you launch Jupyter and Spyder or create new environments with a couple of clicks.

That's it. Once installed you are ready to write code.

Launch Jupyter Notebook

Once Anaconda is installed, you're just a few clicks away from your first interactive workspace. Open Anaconda Navigator and click Launch under Jupyter Notebook. Your default browser will open a new tab showing the Jupyter file browser; choose or create a folder where you want to store your work. From there, select New > Python 3 and Jupyter will spin up a fresh notebook, an empty canvas made of code cells and narrative text blocks, as shown in Figure 1-1.

Notebooks shine because they let you run code one cell at a time, inspect the output immediately, tweak your logic, and rerun on the spot—a perfect loop for iterative math modeling. You can interleave Markdown notes, equations, and plots right beside the code that generates them, turning each notebook into a living, self-documenting analysis. And because everything happens inside your browser, what you see onscreen should match the screenshots in this book line for line.

Cloud-friendly Alternatives

If the security settings on your machine don't allow installation of new software or you simply want a zero-setup option, a browser-based notebook is the fastest path forward. Google Colab (`colab.research.google.com`) is the most popular choice. Sign in with any Google account and click New Notebook. You'll instantly be coding on Google's servers with free access to GPUs and most of the common data-science libraries preinstalled. Your work is saved to Google Drive, so you can pick up exactly where you left off from any device, and sharing a live notebook with a colleague is as easy as sending a link. See Figure 1-2.

Another option is Anaconda Notebooks (`www.anaconda.com/products/notebooks`), a cloud workspace maintained by the same team behind the desktop distribution. It provides the familiar Jupyter interface, complete with conda environments and the full Anaconda package catalog, without requiring anything on your local machine. Projects live in the cloud, updates happen behind the scenes, and collaborative editing feels no different from working in a shared document.

Both platforms free you from installation headaches and make collaboration painless: open a browser, open a notebook, and it just works. Jupyter Notebook, Google Colab, and Anaconda Notebooks all share the same familiar structure.

Now that you can work with Python, let's dive deeper into the ecosystem.

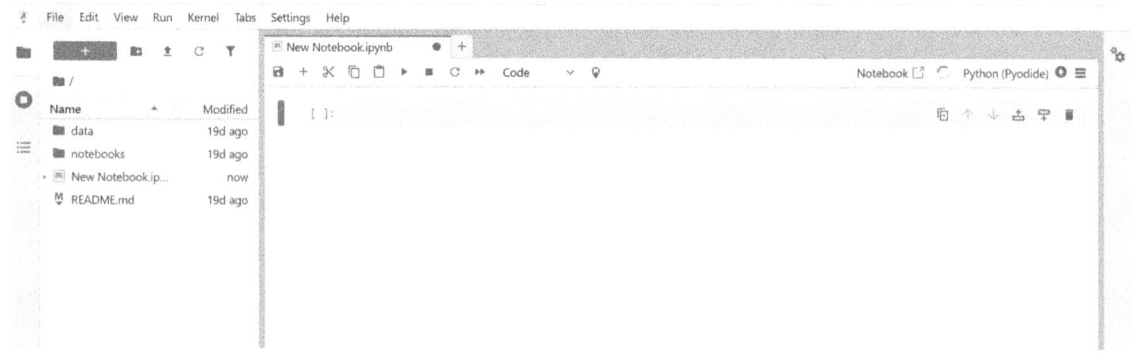

FIGURE 1-1: Jupyter Notebook.

FIGURE 1-2: Google Colab Notebook.

THE PYTHON ECOSYSTEM

Python thrives because of its robust ecosystem of open-source libraries, collections of pre-written code created and maintained by the global Python community. These libraries make it easy to perform complex tasks without writing everything from scratch. At the time of writing in this book, the latest version of Python is 3.14.

For business applications, several libraries are especially important:

➤ **Pandas:** This is the go-to tool for manipulating tabular data (think Excel spreadsheets). It provides data structures like DataFrames that allow you to filter, group, join, and summarize datasets with ease.

➤ **NumPy:** Short for Numerical Python, NumPy enables efficient numerical computations. It is the foundational package for scientific computing in Python and underlies many other libraries like Pandas and scikit-learn.

➤ **Matplotlib:** One of the oldest and most widely used Python visualization libraries. It provides functions to create line plots, bar charts, histograms, and more. Great for producing publication-quality static graphics.

➤ **Seaborn:** Built on top of Matplotlib, Seaborn makes it easier to create complex statistical graphics with simpler syntax and more attractive defaults.

➤ **scikit-learn:** A powerful library for machine learning. With just a few lines of code, you can run clustering algorithms, build classification models, or perform regression analysis.

➤ **Plotly:** A library for creating interactive visualizations and dashboards. Often used for business presentations, it enables dynamic charts that users can explore in real time.

> **NOTE** You'll find the link to the package documentation in this chapter's closing section, "Continue Your Learning."

The modularity feature makes Python incredibly efficient for business tasks. Whether you're building a financial model, cleaning customer data, or visualizing a product launch's performance, Python's ecosystem has you covered. New libraries appear almost weekly, and they're battle-tested by thousands of practitioners before you even download them.

What Is a (Jupyter) Notebook?

A notebook is both scratch pad for your Python ideas and final presentation to distribute your final code. It makes Python friendly for anyone who needs transparent, reproducible analysis.

You begin by opening a Jupyter Notebook in your web browser. The notebook is a blank page made of "cells." Each cell can be one of two types—code or Markdown cells. A *code cell* holds Python statements; press Shift+Enter and the code runs, with the result appearing directly underneath. A *Markdown cell* holds notes, headings, or equations written in plain text. By alternating code and Markdown, you build a step-by-step narrative that records every assumption, calculation, and conclusion in the order they happen.

This structure has practical benefits. First, the immediate feedback loop lets you test an idea, see the outcome, and adjust it without leaving the page. Second, the finished notebook is self-documenting— anyone can scroll top-to-bottom and understand what you did and why, without digging through hidden formulas. Third, sharing is easy. You can send the `.ipynb` file for colleagues to rerun or export it as HTML or PDF for a read-only report. Some notebooks even have the option to share a direct link for collaboration.

Installing Libraries Locally or in a Notebook

Most of the libraries discussed earlier and used in this book (Pandas, NumPy, and Matplotlib) come preloaded in either Anaconda or Google Colab, so you can jump straight into the examples without touching a package manager. Eventually, you will want something that is not in the starter pack (maybe *yfinance* for stock data or *prophet* for time-series forecasting). Adding a new library takes less than a minute, and you can do it in two ways.

If you're working on your own machine with Anaconda, open an Anaconda prompt (Windows) or a terminal (macOS/Linux) and run the following:

```
conda install library-name
```

Conda will resolve dependencies, ask for confirmation, and wire everything up. If the package isn't available through conda, fall back to Python's built-in installer:

```
pip install library-name
```

The new code is immediately available the next time you start a notebook or script.

When you're inside a notebook (Jupyter or Google), you don't even have to leave the page. Instead, you can prepend an exclamation point to the same command and run it in its own cell:

```
!pip install library-name -quiet
```

The notebook will stream the install log right in the output area. Once the prompt returns, you simply import the library into a new cell and keep going. If the kernel was already using that library in the background, a quick Kernel > Restart (Jupyter) or Runtime > Restart Runtime (Colab) command reloads the fresh version.

That's all there is to it: one command in a terminal when you're local, one command in a cell when you're in the browser. No hunting for ZIP files, no admin privileges, and no wait time between deciding you need a tool and putting it to work.

WRITING YOUR FIRST PYTHON SCRIPT

If you've never run Python before, one of the quickest ways to see it interact with the real world is to display today's date. Open a new notebook cell (in Jupyter or Google Colab), type the following three lines, and press Shift+Enter:

```
#Calculate margins and growth rates
from datetime import date
today = date.today()
print("Today is:", today)
```

Here's what each line does:

➤ **Line 1:** `from datetime import date`

Python comes with a built-in standard library, a collection of ready-made tools. The `datetime` module is one of those tools, and inside it lives a helper class called `date`. By writing this `import` statement, you tell Python, "Please give me the date helper so I can use it."

➤ **Line 2:** `today = date.today()`

Now that the helper is available, you call its `today()` method, which asks your computer (or Colab's server) for the current calendar date. Whatever comes back (e.g., 2025-06-30) is stored in a variable named `today`. A variable is just a labeled box that keeps a value in memory so you can reuse it later.

➤ **Line 3:** `print("Today is:", today)`

Finally, you display the result. The `print()` function sends whatever you give it to the screen. Here you pass two things: the text `"Today is:"` and the value stored in `today`. Python stitches them together and shows the message in the notebook output area.

Run the cell and you'll see something like:

```
Today is: 2025-06-30
```

That's your first working Python program. Three simple lines: import a tool, capture a value, and return the result.

SUMMARY

This chapter laid the groundwork for your journey into Python-powered business analytics. The chapter started by exploring why Python has evolved from a simple scripting language to an essential tool for modern business leaders. You learned that while spreadsheets are excellent for quick tasks, they often hit a wall with large datasets and complex scenarios. Python offers the scalability to handle millions of rows and the agility to automate tedious workflows, allowing you to move from manual reporting to automated insight.

This chapter also demystified the technical setup, ensuring you have a professional-grade environment ready to go. You learned how to install the Anaconda distribution for a robust local setup and how to utilize cloud-based alternatives like Google Colab for immediate access. I introduced the Jupyter Notebook as your primary workspace, a powerful interface that combines live code, visualizations, and narrative text into a single, shareable document. Finally, you wrote your first Python script, a simple but symbolic first step. You are now equipped with the tools, the environment, and the understanding needed to tackle the advanced mathematical and analytical challenges in the chapters ahead.

CONTINUE YOUR LEARNING

The following are some resources you may find useful when getting started with Python:

➤ Welcome to Colab: Getting started with Google Colab: `https://colab.research.google.com`

➤ Anaconda Notebooks: `https://www.anaconda.com/docs/tools/anaconda-notebooks/main`

➤ Getting Started with Anaconda: `https://www.anaconda.com/docs/getting-started/getting-started`

➤ Jupyter Notebook Quick Start: `https://jupyter-notebook.readthedocs.io`

➤ Python Official Tutorial: `https://docs.python.org/3/tutorial`

In addition, the following are links to the documentation for the most commonly used math-related Python packages:

➤ **Pandas:** `https://pandas.pydata.org`

➤ **NumPy:** `https://numpy.org`

➤ **Matplotlib:** `https://matplotlib.org`

➤ **Seaborn:** `https://seaborn.pydata.org`

➤ **scikit-learn:** `https://scikit-learn.org`

➤ **Plotly:** `https://plotly.com`

2

Basic Mathematical Operations in Python

Whether you are forecasting next quarter's revenue, dissecting gross-margin trends, or experimenting with a new pricing curve, every business calculation rests on a bed of basic arithmetic and algebra. The stronger that foundation, the easier it is to adjust assumptions when markets shift, or you are presented with new data. This chapter explores how Python does math.

NUMBERS, VARIABLES, AND FUNCTIONS: THE FOUNDATIONS OF BUSINESS LOGIC

Accurate math underpins almost every quantitative decision, from forecasting revenue and allocating budgets to setting prices or stress-testing assumptions. To perform those operations reliably in Python, two concepts must be clear at the outset: variables and data types.

A *variable* is simply a name that points to a value in memory. Assigning a number or string to a variable allows you to preform mathematical operations naturally. Unlike other programming languages, Python lets you reassign a variable, even if the new value has a different data type. The assigned *data type* determines how arithmetic or comparisons in the code will behave.

When you treat everything as "just a number," subtle errors creep in: add the text "3" to the integer 3 and you get string concatenation instead of arithmetic addition. If you have ever mixed a string and a number in Excel and been rewarded with #VALUE!, you have already seen why data types matter. Python enforces these distinctions more predictably than a spreadsheet, but only if you stay aware of what each variable contains.

This chapter explores how Python represents numbers, text, and logical values, and how to perform mathematical operations with them. It begins by examining variables and their underlying types. Gaining clarity on these basics now will help you avoid subtle but costly mistakes as the math becomes more complex.

Understanding Variables

Each variable has two parts: the name you choose and its corresponding value. That value can be a number, a string, or even the result of another calculation. In the following example, five different variables are assigned. Notice how well named labels transform code from a cryptic formula into a readable sentence:

```
unit_price   = 25
units_sold   = 1_000
total_sales  = unit_price * units_sold

discount_rate = 0.1
net_sales     = total_sales * (1 - discount_rate)
print(net_sales)
```

The output from running this code is as follows:

```
22500.0
```

What you name your variables is up to you, but a few conventions will keep the code readable:

- ➤ **Choose descriptive names:** `units_sold` is cleaner than `x`; `discount_rate` beats `dr`. The extra keystrokes pay for themselves when you return to the script six months later.

- ➤ **Use snake_case:** Lowercase words separated by underscores (`gross_margin`) are the de facto Python style and avoid the spaces Excel allows.

- ➤ **Insert underscores in long numbers:** `1_000_000` is far easier to scan than `1000000`, and Python ignores the underscores at runtime.

- ➤ **One concept, one variable:** Resist the urge to recycle names. If profit later becomes *adjusted* profit, give it a fresh label such as `adj_profit` and keep both values side by side.

- ➤ **Keep units obvious:** Add a suffix if it prevents confusion: `price_usd`, `weight_kg`, `rate_pct`.

- ➤ **Plural for collections, singular for scalars:** `regions = [...]` signals a list, while `region = "North"` signals a single value.

- ➤ **Avoid Python keywords:** Names like `list`, `sum`, and `id` are reserved for built-in Python functions and will cause subtle bugs.

Because variables are stored using plain text in your file, you can change the value stored in `unit_price` once and every downstream calculation is updated, reducing the need for manual search-and-replace. This makes Python ideal for "what-if" modeling: tweak a single assumption, rerun, and see the impact instantly.

In short, numbers and variables in Python work the way you would expect. By following these naming conventions, your Python scripts will remain clear, readable, and maintainable, enhancing their value as reliable decision-making tools.

Arithmetic in Python

In Python, arithmetic works just like on a calculator. Any value you type without quotes becomes a number, and you can combine numbers with the six core operators.

```
# Example: Profit calculation
revenue = 150000
expenses = 95000
profit = revenue - expenses
profit_margin = profit / revenue
print(profit_margin)
```

In this example, two variables store the company's revenue and expenses. Subtracting expenses from revenue gives the profit, which is then divided by revenue to calculate the profit margin. The output from running this code is as follows:

```
0.36666666666666664
```

This is Python's exact decimal output; later you'll learn how to format numbers for readability. You can take the same approach for any calculation you want to perform in Python. Table 2-1 shows the six core operators.

TABLE 2-1: The Core Operators in Python

OPERATOR	OPERATION
+	Addition
−	Subtraction
*	Multiplication
/	Division
**	Exponentiation*
%	Modulus

* It's important to note: The use of * for exponentiation is different from either the ^ or the regular exponent key you may be used to. You may get back a result, but it won't be a number raised to its exponent.

> **NOTE** *The modulus operator returns the remainder after division. For example: 3% 1 returns 0.*

Most mathematical operations in Python work as you would expect them to. Exponentiation, however, is a little different: Python uses `**` instead of the caret `^` symbol you may be used to from spreadsheets or calculators. Finally, Python observes the usual PEMDAS order of operations, so parentheses are essential for making a complex expression evaluate exactly as intended.

Working with the *math* Module

So far, we have relied on Python's built-in arithmetic operators to add, subtract, multiply, and divide. When you need functions beyond the basics (square roots, logarithms, trigonometry, precise rounding), the standard-library `math` module is a good starting point. It operates on single numbers (scalars), is available in every Python installation, and executes in efficient C under the hood. To include the `math` module in your Python programs, you use an `import` statement:

```
import math
```

Once you have imported the `math` library, you can use additional functions such as, `math.sqrt(x)`, `math.exp(x)`, `math.log(x, base)`, or even constants such as `math.pi`.

Most functions in the `math` module carry out one operation at a time. If you need to carry out calculations on thousands of numbers, calling `math.sqrt` in a loop will feel sluggish. In those cases, you should use the NumPy library, which performs the same operation on an entire array in one pass. You learn more about NumPy later in this chapter.

DATA TYPES IN PYTHON

While operators and math functions let you perform calculations, the way Python interprets those calculations depends on the type of data you're working with. Every number, piece of text, or logical value in Python has a type. Because Python is *dynamically typed*, you don't need to declare these types explicitly, but understanding them is key to avoiding unexpected results.

Core Data Types

This section introduces Python's built-in data types, which sets a foundation for future mathematical calculations. It starts with numbers and then moves on to text, Booleans, and the special `None` value for missing data.

Numbers: int and float

An *integer* (*int*) is a whole number, ideal for anything you count rather than measure: units sold, headcount, invoice quantity, the number of weeks in a quarter. Python stores integers with arbitrary precision, which means you never risk overflow when totals climb into the millions or even billions. In the following, 25 is interpreted as an int:

```
units_sold = 25
```

A *float* represents a real number with a fractional component, for example, prices, tax rates, interest percentages, or conversion factors:

```
unit_cost = 2.55
profit_margin = 0.33
```

Whenever a calculation involves at least one float, Python automatically "promotes" the other value to a float so that the result preserves the decimal information. Consider an example based on calculating profit. If you're calculating profit based on units sold and unit cost, then your unit count is likely an int, but your unit cost is a float. If a coffee roaster sells 25 bags of beans at $2.55 each and wants to see the gross profit and margin, then the bag count is an int and the price is a float. Python mixes the two data types seamlessly:

```
units_sold    = 25
unit_cost     = 2.55
selling_price = 4.00

revenue = units_sold * selling_price
cost    = units_sold * unit_cost
profit  = revenue - cost
margin  = profit / revenue

print(margin)
```

When running this listing, the resulting value of `margin` should be printed:

```
0.3625
```

The calculations return floats whenever a decimal is involved, ensuring precision. Although mixing the two types is effortless, it is worth remembering that they live on different footing. Integers are exact; floats are approximations held in binary form, which means some seemingly simple decimals (0.1 or the decimal form of 1/3) cannot be stored with perfect precision. This can lead to silly rounding errors. For example, try entering 1.1 + 2.2 in the Python terminal command line. For most reporting and analytic tasks, the resulting microscopic error is irrelevant, but not in workflows that settle money to the cent. Knowing which values are int versus float matters later when you format currency or round percentages for a report. As you would expect, the value $36.25 looks cleaner than 36.249999999.

Text: str

A *string (str)* is any ordered sequence of characters, whether letters, digits, punctuation, or spaces. In day-to-day analysis these sequences carry the labels that make numeric tables intelligible: customer names, product IDs, region codes, status flags. Each of the following is a string:

```
customer_name = "Acme Corporation"
product_id = "SKU1234"
product_upc = "781118823774"
```

Because strings represent qualitative information, they often become the keys by which you group, join, or filter data—*sales by region, returns by product ID, feedback by customer segment.* Unlike numbers, strings preserve case and punctuation exactly as typed, so consistent spelling and capitalization matter when you expect two labels to match.

True/False Logic: bool

Booleans capture binary states, a single bit of information: on/off, yes/no, pass/fail. They are the backbone of conditional logic and filtering operations. In Python, Booleans can be either `True` or `False`. For example:

```
is_active = True
inventory_low = False
```

Booleans are case sensitive. That is, Python will recognize True, but not true as a Boolean. Using Booleans in a Pandas DataFrame lets you pull out only the rows that meet a criterion: *orders over $10 000, customers who opted in, items flagged for restock.* In control structures (`if`, `while`, and list comprehensions), a Boolean determines which branch of code runs, keeping decision rules explicit and easy to audit.

Missing or Null Values: None

Real-world datasets rarely arrive complete. Python's sentinel value `None` signals "no data here yet." The `None` keyword can be stored in a variable just like any other value:

```
delivery_date = None
```

Setting values to `None` as a placeholder can be useful when a variable is used later in the codebase. With Pandas, the `None` value is useful when there is a missing observation or missing data. Treating missing values explicitly prevents hard-to-trace errors when you later add, divide, or plot the data.

Why Data Types Matter

Imagine you're calculating customer lifetime value (CLV), and one of your data sources stores customer tenure as text instead of numbers. A value like "3" (a string) won't behave the same as 3 (an integer). If you try to do math with the wrong type, Python will warn you, or worse, silently give the wrong result, as the following code illustrates:

```
#Incorrect: Concatenates two strings
print("3" + "5")
# Correct: Adds the numbers
print(3 + 5)
```

When you run this code, you will see that the first `print` statement results in `35` being printed while the second results in the value of `8` being printed. Adding `"3"` to `"5"` results in strings being concatenated rather than numbers being added together. This type mismatch (mistaking strings for numbers) happens all the time when importing data from Excel, CSVs, databases, or APIs. Even if something looks like a number, it may be stored as a string. As a result, this is a common source of subtle errors. Ensuring you explicitly handle these conversions early prevents inaccurate calculations and costly mistakes. Being aware of this helps you write safer, more predictable code, and avoids countless hours of debugging.

Converting Between Types

Python gives you a small set of built-in functions that act like directors, telling a value to play a different role when the scene calls for it. These include `int()`, `float()`, `str()`, and `bool()`. Here are a few examples of what each one does and when you'll reach for it:

➤ `int()`: Convert to a whole number

```
int("20")       # 20
int(19.99)      # 19       (truncates the decimal)
```

Notice how the `int()` function handles the floating-point number `19.99`—it does not serve as a rounding function, but rather extracts the integer part of the number and discards the fractional part. You can use this function with strings when a CSV delivers data such as "units_sold" as text.

➤ `float()`: Convert to a decimal

```
float("19.99")  # 19.99
float(5)        # 5.0
```

This function is helpful whenever percentages or currency arrive as integers and you need the fractional precision for calculations. Again, here, take note of the way 19.99 is now stored.

➤ `str()`: Convert to text

```
str(2024)       # "2024"
str(9.5)        # "9.5"
```

This function is ideal for labels such as `"Q" + str(3)`, which results in `"Q3"`, or for exporting numbers back to a text-based report.

➤ `bool()`: Convert to `True`/`False`

```
bool(1)         # True
bool(0)         # False
```

This function converts many "presence/absence" signals into a single binary column you can filter on.

Think of these functions like changing a cell's format in Excel: the underlying information stays the same, but Python now knows how to treat it. Master these conversions early and you skip a whole class of sneaky bugs, leaving you free to focus on the insights that move the business.

> **TIP** Being mindful of types from the outset keeps the focus on insight rather than troubleshooting.

BUSINESS DATA STRUCTURES: ARRAYS AND MATRICES

Business data often comes naturally organized as arrays and matrices—think financial projections, sales across multiple regions, or quarterly inventory levels. Using NumPy, Python provides efficient tools to represent and manipulate this structured data, significantly simplifying complex analyses.

Before you dive into the code, pause for the mathematics behind it. A *vector* is simply an ordered list of numbers. For example, the list can include revenue for each quarter, daily temperatures, or even three coordinates of a point in space. Stack several vectors side-by-side and you have a *matrix*, the workhorse of linear algebra. Matrices let you rotate a point, solve a system of equations, or encode every scenario of a cash-flow model in a single object. The beauty of treating data as vectors and matrices is that many business calculations can be expressed as compact algebraic rules rather than repetitive scalar arithmetic.

Python's built-in lists can store these sequences, but they don't understand the algebra. Simple addition using the plus operator (+) concatenates them; multiplying two lists raises an error. This is the problem solved by NumPy. NumPy can store numbers in a contiguous block of memory (much like a traditional vector in C or Fortran) and repurposes the arithmetic operators to perform element-wise or matrix operations that mirror the math you learned on paper. Write `price * quantity` when both are arrays and NumPy forms a new array of line-by-line products. Likewise, the @ operator carries out true matrix multiplication, by making expressions such as `weights @ returns`.

This vectorized style boosts productivity. NumPy runs in low-level, compiled code. Its native parallelism and avoids looping when possible. Speed matters when you escalate from five rows to five million, but clarity matters too. By learning to think in vectors and matrices, and by implementing that thinking in NumPy, you gain both mathematical precision and computational efficiency.

The next section explains how arrays and matrices work in more detail, starting with one-dimensional arrays and then expanding to matrices.

One-dimensional Arrays

A vector $s = [s_1, s_2, s_3, s_4]$ can represent four quarters of sales $s = [10000, 12000, 9500, 11000]$. Multiplying that vector by a scalar 1.05 is the textbook definition of scalar-vector multiplication: every component grows by 5%. This is shown in the following code:

```
import numpy as np
sales = np.array([10_000, 12_000, 9_500, 11_000])
# apply 5 % increase
uplift = sales * 1.05
print(uplift)

#Result: [10500. 12600.  9975. 11550.]
```

Within this code, NumPy is imported and nicknamed np (for brevity). NumPy treats `sales` as a single object, which is assigned an array of four values, one for each quarter. These values are multiplied by 105% to produce another array called `uplift` that is of equal length. This is done without using an explicit loop. This can be seen by printing the values of `uplift`, which results in the following output:

```
[10500. 12600. 9975. 11550.]
```

Common vector summaries map directly to statistical definitions:

```
print('Total Sales: ',sales.sum())
print('Average Sales: ',sales.mean())
print('Std. of Sales: ',sales.std())the vector
```

Each line prints the summary statistic of the sales array. When you run these lines, you should see the following results:

```
Total Sales:  42500
Average Sales:  10625.0
Std. of Sales:  960.143218483576
```

Not only can you set the values of an array, but you can generate them as well. Generated sequences are created as follows:

```
weeks = np.arange(1, 53)
rates = np.linspace(0.01, 0.12, 12)
```

In the code, the first line creates an array with 1 row and 52 columns (remember, in Python the columns stop at the number before the one you designate, like in the standard range). The second line creates an array with 12 columns from 0.01 to 0.12. These are not Python loops; NumPy creates the entire vector in one call.

Matrices: Two-dimensional Arrays

A *matrix* is a rectangular grid of numbers. NumPy stores it as a two-dimensional array whose shape is written "rows × columns." The basic arithmetic (sums, means, element-wise addition) is straightforward, but two additional rules matter:

➤ **Dot product (row · column)** generates a single number.

➤ **Matrix multiplication** stacks many dot products so shapes must align: $(m \times n)(n \times p) \to (m \times p)$.

Imagine a retail chain tracking unit costs for multiple products across various regions to optimize pricing. Matrix operations effortlessly aggregate data by product or region, streamlining profitability analyses.

Aggregations by Axis

Let

$$Costs = \begin{bmatrix} 10 & 12 & 9 \\ 14 & 11 & 13 \end{bmatrix}$$

represent unit costs for two products (rows) across three regions (columns).

```
import numpy as np

cost = np.array([[10, 12,  9],
                 [14, 11, 13]])

row_totals = cost.sum(axis=1)
col_means  = cost.mean(axis=0)
```

`axis=1` collapses each row, giving product-level totals; `axis=0` collapses each column, giving region averages. A way to remember this is that zero moves vertically, or down the rows of each column, whereas one moves across or horizontally.

Matrix Multiplication

Suppose unit sales are

$$Sales = \begin{bmatrix} 120 & 150 & 90 \\ 100 & 130 & 110 \end{bmatrix}$$

To obtain revenue, you need one dot product per product–region pair. NumPy handles this if you multiply `Sales` by the transpose of `Costs`, which flips its rows and columns so the inner dimensions match $(2 \times 3) \cdot (3 \times 2)$.

```
# Make sure to run the code under the heading "Aggregations by Axis"
before this snippet.

units   = np.array([[120, 150,  90],
                    [100, 130, 110]])

revenue = units @ cost.T

print(revenue)
```

The resulting matrix is as follows:

```
[[3810 4500]
 [3550 4260]]
```

Each entry in the resulting matrix is formed through standard matrix multiplication. For example, for the upper-left entry of the new matrix (index 1, 1), revenue is computed as revenue(1,1) = 120 × 10 + 150 × 12 + 90 × 9.

NumPy performs all four row–column dot products in compiled code and returns a 2×2 matrix: rows are products, columns are regions.

Broadcasting a Vector

When two arrays do not line up dimension-for-dimension, NumPy virtually "stretches" any dimension of size 1 to match the other array. For example, imagine you have a vector of surcharges to apply across all unit costs (a 2 × 3 matrix).

$$\sigma = \begin{bmatrix} 0.5, 0.7, 0.4 \end{bmatrix}$$

```
surcharge = np.array([0.5, 0.7, 0.4])
landed    = cost + surcharge        # same shape as cost
```

This code snippet adds the surcharge to every row of the matrix `Costs`. NumPy's broadcasting automatically stretches the 1D array across the rows:

Element-wise operations require no loops; NumPy aligns shapes and applies the arithmetic to every entry.

Selection: Slicing and Boolean Masks

Vectors and matrices support slice notation, which lets you select specific rows, columns, or sub-arrays:

```
west      = cost[:, 2]
product_b = cost[1, :]
```

In the first line, `cost[:, 2]` means "take all rows (:) from the third column (2 since indexing starts at 0)." The result is a one-dimensional vector containing the costs for the West region across all products. In the second line, `cost[1, :]` means "take the second row (1) across all columns (:)." The result is the full set of costs for product B across all regions.

Slicing gives you direct access to rows and columns, but sometimes you want data based on a condition, not a position. For that, you can use a *mask*.

Boolean masks act like algebraic indicator functions `I(condition)`, marking which elements meet a condition. For example:

```
high = cost > 12
cost[high]                  # elements where cost_ij > 12
```

The first line creates a condition (or mask), where each entry is either `True` (if the corresponding element is greater than 12) or `False` (otherwise). The second line then uses that mask to return only the elements where the condition holds, effectively filtering the array down to the values above 12. Booleans behave like ones and zeros in arithmetic. You can also use them to compute quick counts, for example, calling `high.sum()` will give you a quick count of the number of costs above 12.

Example: Random Vectors and Monte Carlo

Deterministic models tell you what happens when the input is fixed. In practice, key drivers, demand, lead times, and FX rates bounce around. Monte Carlo simulation tackles that uncertainty by treating inputs as probability distributions rather than single values, then sampling from those distributions many times to form a cloud of possible futures.

Businesses frequently face uncertainty from fluctuating demand, changing interest rates, or uncertain supply-chain lead times. Monte Carlo simulation provides a structured way to assess such uncertainties, making it invaluable for risk management and strategic planning.

Suppose historical data suggests that monthly demand for a product is roughly bell-shaped with a mean of 500 units and a standard deviation of 35. In statistical notation that's $N(\mu = 500, \sigma = 35)$. A single draw from this distribution represents one plausible month; 10,000 draws approximate the full range you might encounter in a year of day-to-day operations.

```python
import numpy as np

rng     = np.random.default_rng(2025)
demand  = rng.normal(loc=500, scale=35, size=10_000)
```

The result, `demand`, is a NumPy vector of length 10,000. Because it is an array, you can apply arithmetic to all scenarios in one step. If the selling price is $12.75, revenue for every simulated month is as follows:

```python
revenue = demand * 12.75
```

Now you have 10,000 revenue outcomes, an empirical distribution. Summaries are immediate:

```python
print("Mean: ",revenue.mean())
print("5th percentile: ", np.quantile(revenue, 0.05))
print("95th percentile: ", np.quantile(revenue, 0.95))
```

The result of running this code together is as follows:

```
Mean:  6376.773217565064
5th percentile:  5628.634343409629
95th percentile:  7107.219698663389
```

The following code helps visualize the code by using another popular library, matplotlib. The results of running this code are shown in the graph in Figure 2-1.

```python
import matplotlib.pyplot as plt

plt.hist(revenue, bins='auto')
plt.title("Monte-Carlo Histogram")
plt.show()
```

Plotting a histogram shows the shape of potential results; computing the 5% tail quantile gives a back-of-the-envelope risk measure. All of this flows from straightforward code. Vectorization keeps the code short, and NumPy's underlying C routines keep it fast, which is essential when you scale the simulation to multiple products, correlated drivers, or tens of thousands of scenarios.

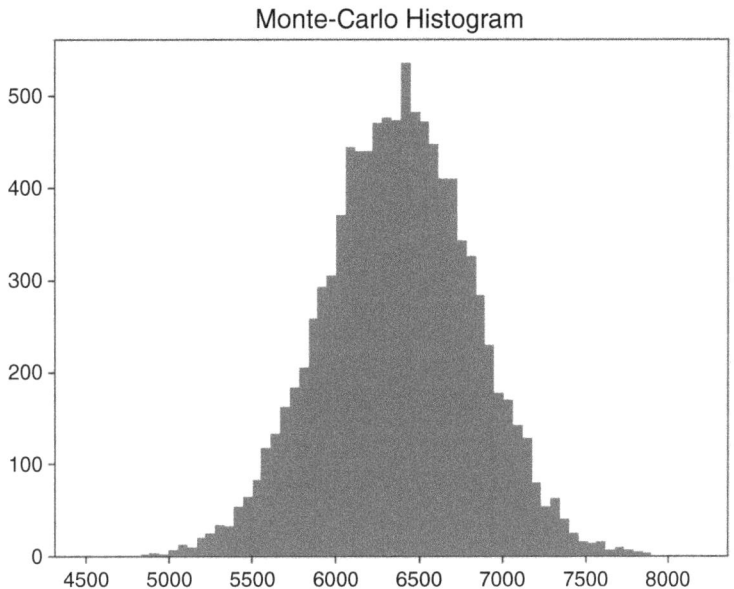

FIGURE 2-1: A histogram produced by matplotlib.

In summary, two-dimensional arrays let you express row/column summaries, dot products, and full matrix products with the same concise syntax you'd write on paper, executed at machine speed.

DATA MANIPULATION BASICS WITH PANDAS

A raw data file (CSV, Excel export, or database dump) rarely answers a question outright. Fields need renaming, totals need adding, and subsets need isolation before a single chart or model makes sense. *Pandas builds on the foundation of NumPy,* but provides a more general data manipulation toolkit. A DataFrame acts like an in-memory spreadsheet that you control entirely through code: you can slice rows, pick columns, compute new metrics, or reshape the grid for a different perspective, all without leaving the Python prompt. The next sections walk a typical path from loading data to producing a clean summary that can feed visualizations, dashboards, or downstream modeling.

Constructing a DataFrame

Data arrives in many forms but Pandas treats it all the same way: as rows and columns. That abstraction lets you build a table from a Python dictionary, a CSV file, or a SQL result set with just one call.

```
import pandas as pd

# From a dictionary ------------------------------------------------
sales = pd.DataFrame({
    "Region": ["North", "South", "East", "West"],
    "Month" : ["2025-01", "2025-01", "2025-01", "2025-01"],
    "Revenue": [15_200, 18_100, 12_900, 17_600]
})
```

```
# From a CSV file ----------------------------------------------
orders = pd.read_csv("https://raw.githubusercontent.com/bkrayfield/
Applied-Math-With-Python/refs/heads/main/Data/orders_2025_Q1.csv")
```

Helper functions like `DataFrame.shape` can report rows × columns, while *dtypes* shows each column's type.

First Looks: *head(), info(), describe()*

Pandas gives you a simple way to investigate your data. These helper functions help you investigate your dataset:

➤ `head()` (similar to the UNIX `head` command) gives a sample of the first few rows of the DataFrame.

➤ `info()` lists dtypes and non-null counts: a structural sanity check.

➤ `describe()` computes column-wise reductions (mean, std, quartiles) in one call, using the same reduction kernels that NumPy applies to an ndarray.

For example, when you call `.head()`, you get the output shown in Figure 2-2.

These functions are read-only; they leave the DataFrame unchanged. These three commands form a quick triage: Are there missing values? Do column types match expectations? Are any metrics wildly off?

Working with Columns and Rows

A column behaves like a named list; a row like a record.

```
# Column access
revenue_series = sales["Revenue"]
subset         = sales[["Region", "Revenue"]]
```

The first line selects just the Revenue column from the `sales` DataFrame and stores it as a one-dimensional series. The second line selects two columns at once, Region and Revenue, returning a smaller DataFrame with only those fields.

	OrderID	Region	Month	Units	UnitCost	Revenue
0	10001	North	2025-01	200	9.50	1900
1	10002	South	2025-01	240	9.25	2220
2	10003	East	2025-01	180	8.90	1602
3	10004	West	2025-01	230	9.30	2139
4	10005	North	2025-02	210	9.60	2016

FIGURE 2-2: Output from calling .head().

Rows can be selected by number (`iloc`) or by label (`loc`):

```
# Row labels vs. integer position
north_row = sales.loc[0]          # label-based (index value)
second_row = sales.iloc[1]        # position-based
```

Here, `sales.loc[0]` fetches the row with index label 0, while `sales.iloc[1]` fetches the second row in order. Using `loc` vs. `iloc` makes it explicit whether you're referring to labels or numeric positions, which is helpful for avoiding the off-by-one mistakes common in spreadsheets.

Filtering with Booleans

The expression `sales["Revenue"] > 17_000` broadcasts the comparison down the column, returning a series of `True`/`False` values, one for each row. That Boolean series can then be used as a filter.

```
high_rev = sales[sales["Revenue"] > 17_000]
january  = sales[sales["Month"] == "2025-01"]
```

In the first line, only rows with Revenue greater than 17,000 are returned. In the second, only the rows where the Month equals 2025-01 are selected. Both create new filtered DataFrames, leaving the original untouched.

You can also combine multiple conditions using & (and), | (or), and parentheses:

```
north_jan = sales[
    (sales["Region"] == "North") &
    (sales["Month"]  == "2025-01")
]
```

Here, two conditions are combined: Region must equal North and Month must equal 2025-01. The result is a subset containing only January sales from the North region.

Creating New Columns

You can create new columns in a variety of ways, including providing a new list of values, or even providing a calculated function. Adding or subtracting a scalar from a column is the same element-wise vector operation that NumPy performs:

```
sales["Cost"]   = [9_400, 10_300, 8_100,  9_900]
sales["Profit"] = sales["Revenue"] - sales["Cost"]
sales["Margin"] = sales["Profit"]  / sales["Revenue"]
```

The first line creates a new column named `Cost` by assigning a list of four numbers, one for each row in the DataFrame. The second and third lines compute both profit and margin by creating new columns for intermediate calculations. The result indicates how much profit each row earns as a fraction of its revenue.

Because these calculations are vectorized, every operation is applied to the full column without loops or manual formulas. This is one of the main advantages of using Pandas over spreadsheets: the logic is expressed once, and it applies everywhere automatically. After creating the new columns, you can immediately use them for further analysis. For example:

```
sales.sort_values("Margin", ascending=False).head()
print(sales.head())
```

The first line sorts the DataFrame by Margin in descending order, so the rows with the highest profit margins appear at the top. The result from calling .head() on the sorted DataFrame is shown in Figure 2-3.

Grouping and Aggregation

Grouping splits the data by a key, applies reductions, and recombines the pieces into a tidy summary. If you're familiar with SQL, the logic is very similar. Perfect for a regional roll-up:

```
region_summary = (
    sales
    .groupby("Region", as_index=False)
    .agg(
        Revenue_sum    = ("Revenue", "sum"),
        Profit_mean    = ("Profit",  "mean"),
        Margin_median  = ("Margin",  "median")
        print(region_summary.head())
    )
)
```

The output is shown in Figure 2-4.

Now you have a readable four-row table, one row per region, ready for a bar chart or a bullet point in the deck.

	Region	Month	Revenue	Cost	Profit	Margin
0	North	2025-01	15200	9400	5800	0.381579
1	South	2025-01	18100	10300	7800	0.430939
2	East	2025-01	12900	8100	4800	0.372093
3	West	2025-01	17600	9900	7700	0.437500

FIGURE 2-3: Output from calling .head() on a sorted DataFrame.

	Region	Revenue_sum	Profit_mean	Margin_median
0	East	12900	4800.0	0.372093
1	North	15200	5800.0	0.381579
2	South	18100	7800.0	0.430939
3	West	17600	7700.0	0.437500

FIGURE 2-4: Output from grouping.

Joins and Merges

Business data rarely lives in one table. Use merge to bring it together on a key field. Suppose your sales targets reside in one file, while actual revenue data resides in another. A quick merge in Pandas immediately reveals performance gaps, allowing management to act swiftly on underperforming regions.

```
targets = pd.read_csv("https://raw.githubusercontent.com/bkrayfield/
Applied-Math-With-Python/refs/heads/main/Data/sales_targets.csv")
sales_with_target = sales.merge(targets, on="Region", how="left")

sales_with_target["Gap"] = (
    sales_with_target["Revenue"] - sales_with_target["Target"]
)
```

The first line reads the `targets` file into a DataFrame. The `merge` statement then joins `sales` and `targets` on the Region column, keeping all rows from the left table (`sales`). Finally, the new Gap column calculates how far revenue is above or below target.

The `how` argument controls whether unmatched keys are kept (`left`, `right`, `outer`) or discarded (`inner`). Table 2-2 shows how Pandas matches keys. Any gaps show up as `NaN`, which you can address explicitly.

TABLE 2-2: Pandas Merging Methods

MERGE TYPE	DESCRIPTION	ROWS KEPT	EXAMPLE USE
inner	Keeps only rows where the key appears in both tables.	Matches only	Comparing data where both sources must overlap (e.g., customers with both orders and payments).
left	Keeps all rows from the left table; fills gaps from the right with NaN.	All from left	Preserving all sales data even if some regions don't have targets.
right	Keeps all rows from the right table; fills gaps from the left with NaN.	All from right	Preserving all target values even if some regions have no sales.
outer	Keeps all rows from both tables; unmatched rows are filled with NaN.	All from both	Auditing to see every region in either dataset, whether or not a match exists.

Reshaping: Pivot, Melt, Stack

Sometimes your data needs to change shape before you can analyze it effectively. Columns may need to become rows or rows may need to become columns. This could happen when preparing a heatmap, building a crosstab, or feeding data into another tool.

```
pivot = sales.pivot(index="Month", columns="Region", values="Revenue")
```

Here, the DataFrame is pivoted so that months run down the rows, regions spread across the columns, and revenue values fill the grid. This wide format makes side-by-side comparisons straightforward.

```
long   = pivot.reset_index().melt(id_vars="Month",
                                  var_name="Region",
                                  value_name="Revenue")
```

The `melt` function reverses the transformation, taking the wide table and collapsing it back into a tall, tidy format. Each row now represents a single observation: the month, the region, and its revenue.

With just a handful of verbs—`select`, `filter`, `compute`, `group`, `join`, and `reshape`—Pandas transforms raw tables into concise views tailored to the question at hand. Because each step is expressed directly in code, the path from data to insight is reproducible, reviewable, and easy to automate.

SUMMARY

This chapter established the foundational skills you need to perform accurate and reliable business calculations in Python. You started by mastering variables, learning that clear, descriptive names are key to writing code that users of your code can understand. You then explored Python's core data types (`integers`, `floats`, `strings`, and `Booleans`) and learned why distinguishing between them is critical for avoiding costly errors in financial models.

From there, the chapter moved beyond scalar math to the powerful world of vectorization. You learned how NumPy arrays allow you to perform operations on entire datasets instantly, replacing slow loops with lightning-fast algebraic syntax. You then saw how to extend this to matrices, enabling you to calculate revenue across multiple products and regions in a single line of code. Finally, the chapter introduced Pandas, the ultimate tool for structured data. You learned to load, filter, group, and reshape complex datasets, turning raw files into clean, actionable insights.

CONTINUE YOUR LEARNING

To further solidify your understanding of mathematical operations, data manipulation, and their applications in Python for business analytics, consider exploring these additional resources.

- ➤ **NumPy documentation:** `numpy.org`
- ➤ **Pandas documentation:** `pandas.pydata.org`
- ➤ **Python data types:** `https://docs.python.org/3/library/datatypes.html`
- ➤ **Math functions of the math module:** `https://docs.python.org/3/library/math.html`

3

Visualization for Business Decision-making

Decision-makers respond not just to raw data but to the way that data is communicated, and visualization is the bridge that transforms tables of figures into actionable insight.
A well-crafted chart can highlight seasonality in sales, reveal inefficiencies across departments, or show how close the business is to hitting annual targets. Just as importantly, an effective visual saves time: executives can grasp a story in seconds that might take pages of spreadsheets to explain.

This chapter explores how Python turns business data into visuals that clarify, persuade, and inform. It begins with the foundations in Matplotlib, where you learn how plots are structured and how to build simple line and bar charts. These basic skills mirror the kinds of graphs you can create in Excel, but with the flexibility to customize every element.

THE LANDSCAPE OF VISUALIZATION TOOLS IN PYTHON

Python has become one of the dominant languages for math and data analysis. One reason for its popularity is its rich ecosystem of visualization libraries. Each tool is designed with a different philosophy in mind. Some libraries prioritize publication-ready static charts, some emphasize interactive exploration, and others make it possible to build fully functional dashboards for decision-making. Understanding the strengths of each library helps a user choose the right tool for the task.

At a high level, most visualization falls into three categories:

➤ **Static charts** for reports, publications, or presentations.

➤ **Interactive exploration** to quickly identify patterns and anomalies.

➤ **Dashboarding and applications** for real-time decision-making and stakeholder engagement.

TABLE 3-1: Libraries for Visualization in Python

LIBRARY	MOST SUITABLE FOR	OUTPUT STYLE
Matplotlib	Static plots, fine-grained customization	Publication-quality images
Plotly	Interactive charts, web dashboards	Web-based (HTML/JS)
HoloViz	App-like data tools, dashboards	Web apps, notebooks

What you hope to achieve should dictate the Python visualization library you choose. By choosing the proper library, you can avoid headaches when adapting your code to the proper output format later. Table 3-1 summarizes three of the most widely used options in Python.

When you begin visualizing data in Python, your journey almost always starts with Matplotlib. It is the foundational library of the entire Python visualization ecosystem, and its concepts influence nearly every other tool. Matplotlib's primary strength is its power to produce high-quality, static, publication-ready charts. It gives you granular control over every single element, from axis scales and line thickness to color palettes and font choices, making it the perfect tool for creating a polished, static report.

When you need your audience to explore the data, you'll turn to an interactive library. Plotly is a library designed to bring data to life with interactivity. Instead of static images, Plotly charts allow users to hover over data points, zoom into regions of interest, and toggle variables on and off. This makes it invaluable for exploratory analysis and for presenting data to stakeholders in an engaging way.

Another powerful approach to interactivity comes from hvPlot, a key part of the HoloViz ecosystem. The power of hvPlot lies in its simplicity; it provides a high-level API that feels just like the `familiar.plot()` method on a pandas DataFrame. With minimal code, hvPlot generates fully interactive charts (using other libraries like Bokeh or Plotly as a backend) that are immediately ready for exploration.

Visualization Applications: Dashboarding Frameworks

Sometimes, a single chart isn't enough. You need to build a complete, app-like dashboard with drop-down menus, sliders, and multiple visualizations that update in response to user input. This is where dashboarding frameworks, which are distinct from the charting libraries, come in.

Both of these interactive libraries has a powerful dashboarding partner:

> ➤ **Plotly** is paired with Dash. While Plotly creates the individual interactive charts, Dash is the companion framework used to assemble those charts into a complete, sophisticated web application. You use Plotly to design the "what" (the chart) and Dash to build the "how" (the surrounding app and its controls).

> ➤ **hvPlot** (and the wider HoloViz ecosystem) is paired with Panel. Panel is the dashboarding framework designed to assemble the interactive charts you create with hvPlot into a coherent application. Furthermore, Panel is "visualization-agnostic," meaning it can also build dashboards using charts from Matplotlib, Plotly, Bokeh, and others. This makes it a uniquely flexible integration tool.

Choosing the Right Visualization Tool for Your Work

The choice of library often comes down to audience and context, as well as the required output format for your data visualization. Here are some tips for selecting the correct library:

> ➤ If your goal is to produce a static, print-ready report with full control over the final look, Matplotlib is usually the best choice.

> ➤ If you love the simplicity of the .plot() API but want instantly interactive charts for exploration, hvPlot is an excellent choice.

> ➤ If you need to deliver an engaging and exploratory experience for non-technical users, the Plotly (for charts) and Dash (for the application) stack offers the richest experience.

> ➤ If you want to prototype interactive business apps quickly, especially by combining charts from different libraries, Panel is a powerful and flexible option.

In practice, many organizations use a combination of these tools: Matplotlib for internal analytics and publication-ready visuals, Plotly for interactive stakeholder meetings, and Panel for deploying lightweight decision-support tools.

The remainder of this chapter focuses on Matplotlib for creating visualizations. However, there are many other tools beyond the three mentioned here for supporting data visualizations in Python. Table 3-2 provides information about popular visualization libraries.

TABLE 3-2: Additional Plotting Libraries in Python

LIBRARY	DESCRIPTION	DOCUMENTATION
Seaborn	Built on top of Matplotlib; specializes in statistical visualization with attractive defaults and simple syntax for common analyses.	`https://seaborn.pyda ta.org`
Altair	Declarative statistical visualization library based on Vega-Lite grammar of graphics; concise syntax; best for clean statistical charts.	`https://altair-viz. github.io`
Pygal	SVG-based charting library; creates interactive, lightweight visuals for embedding in web apps.	`http://www.pygal.orgen/ stable/`

(Continued)

TABLE 3-2: (Continued)

LIBRARY	DESCRIPTION	DOCUMENTATION
VisPy	High-performance interactive visualization powered by OpenGL; best for large or complex datasets.	`https://vispy.org`
plotnine	Grammar-of-graphics–inspired plotting library for Python, modeled after R's ggplot2.	`https://plotnine.org`
Cartopy	Geospatial plotting library for creating maps and visualizations of spatial data (often with Matplotlib).	`https://cartopy.readthed` `ocs.io`

GRAPHING BASICS WITH MATPLOTLIB

Matplotlib is the cornerstone of data visualization in Python. It provides the flexibility to create almost any chart you can imagine, from simple line graphs to highly customized multi-panel figures. While newer libraries have emerged to make plotting more convenient, Matplotlib remains essential because it offers complete control over every visual element and serves as the foundation for many other visualization libraries (Seaborn, pandas plotting, etc.).

Just like most libraries in Python, if Matplotlib is not already installed in your Python environment, you can add it simply with `pip`, `conda`, or with any package manager of your choice. For example, you can run this command:

```
pip install matplotlib
```

With this command, you are instructing `pip` to connect to the Python Package Index (PyPI), which is the official online repository for Python software. `pip` then locates the package named `matplotlib`, downloads its files, automatically figures out and downloads any other libraries that Matplotlib depends on, and finally installs all of them into your active Python environment so you can use them in your code.

Understanding the Structure of a Plot

To use Matplotlib effectively, it helps to understand how it organizes a chart. At its core, Matplotlib follows a figure, axes, elements hierarchy. This design may feel abstract at first, but once you see how it works, it becomes intuitive and powerful.

➤ **Figure:** The overall canvas or window that contains everything. Think of it as a sheet of paper. A figure may contain one or many charts.

➤ **Axes:** A specific area inside the figure where the data is actually drawn. Despite the name, "axes" refers to the plot region as a whole, not just the x-axis or y-axis. A figure can contain multiple axes, allowing you to place several plots side by side or stacked on top of each other.

➤ **Elements:** The building blocks that bring a chart to life—titles, x-axis and y-axis labels, tick marks, legends, and gridlines. These are layered on top of the axes to provide meaning and clarity.

An easy way to understand this hierarchy is to think of preparing a business report: the figure is like the entire page, the axes is the specific chart you place in the middle of that page, and the elements are the annotations (headings, captions, and labels) that explain the chart to your audience.

Creating and Working with Plots

The simplest way to create a chart in Matplotlib is to call a plotting function directly. The following code snippet creates the simple plot shown in Figure 3-1:

```
import matplotlib.pyplot as plt

plt.plot([1, 2, 3], [4, 5, 6])
plt.show()
```

This code first imports Matplotlib's plotting module with the common shorthand `plt`. The `plt.plot([1, 2, 3], [4, 5, 6])` command then creates a line chart. Finally, the `plt.show()` function displays the chart in a window or notebook output. Although this produces only a very simple line, it illustrates the basic workflow of Matplotlib: supply data, plot it, and then render the visualization.

The code uses Matplotlib's "state-based" interface, which automatically creates a default figure and axes behind the scenes. It's convenient for quick plots, but when building more complex visuals, it's better to work explicitly with the object-oriented interface. For example:

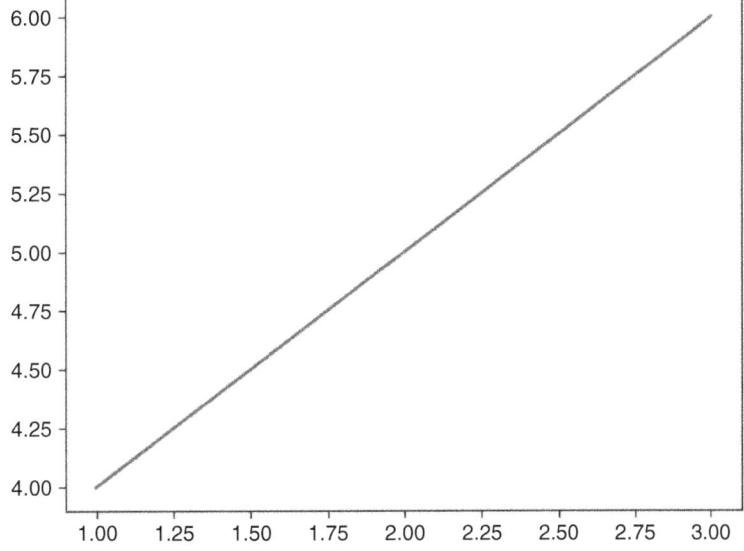

FIGURE 3-1: A simple plot produced by Matplotlib.

```
fig, ax = plt.subplots()
ax.plot([1, 2, 3], [4, 5, 6])
plt.show()
```

In this example, the code uses Matplotlib's object-oriented interface, which provides greater control over plots. The line `fig, ax = plt.subplots()` explicitly creates a figure as the overall canvas and an axes (ax) as the specific plotting area. Next, `ax.plot([1, 2, 3], [4, 5, 6])` draws a line chart on that axes, connecting the points (1,4), (2,5), and (3,6).

Running this code will produce, to the naked eye, the exact same output as Figure 3-1. However, while the output looks similar to the previous example, this style is more flexible and is the recommended approach when you need to add multiple plots, customize chart elements, or manage complex figures.

The example in Listing 3-1 begins to take real advantage of the subplot method. The results of running this code are shown in Figure 3-2.

LISTING 3-1: PLOTTING LINES

```
import matplotlib.pyplot as plt

months = ["Jan", "Feb", "Mar", "Apr", "May"]
revenue = [40000, 45000, 47000, 52000, 55000]
expenses = [30000, 32000, 35000, 37000, 39000]

fig, (ax1, ax2) = plt.subplots(1, 2, figsize=(10, 4))

ax1.plot(months, revenue, marker="o", color="green")
ax1.set_title("Monthly Revenue")
ax1.set_xlabel("Month")
ax1.set_ylabel("Revenue ($)")

ax2.plot(months, expenses, marker="o", color="red")
ax2.set_title("Monthly Expenses")
ax2.set_xlabel("Month")
ax2.set_ylabel("Expenses ($)")

plt.tight_layout()
plt.show()
```

This code snippet has a lot going on. The code begins by setting up the data to be used in the plots. The line `fig, (ax1, ax2) = plt.subplots(1, 2, figsize=(10, 4))` creates a figure and a grid of subplots (which Matplotlib calls *axes*). Let's break down its parameters:

➤ The first two parameters, 1 and 2, define the grid's dimensions. The code is asking for one row and two columns, which will result in two plots arranged side-by-side.

➤ The `figsize=(10, 4)` parameter is a keyword argument that sets the dimensions of the entire figure (the "page" holding the plots) to be 10 inches wide by 4 inches tall.

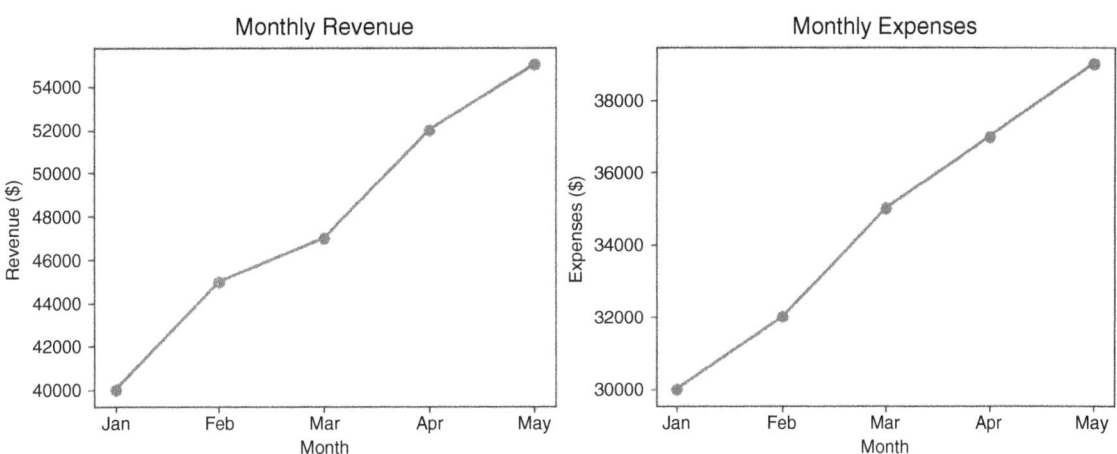

FIGURE 3-2: A 1 × 2 grid of line plots produced using the subplots feature of Matplotlib.

➤ This function returns two types of objects that are "unpacked" into variables. The `fig` variable holds a reference to the overall figure. The tuple (`ax1, ax2`) holds the individual axes objects—`ax1` is the first plot (left) and `ax2` is the second plot (right).

Once the figure and axes are created, you can "draw" on each axes individually. For the first plot, `ax1.plot(months, revenue, marker="o," color="green")` plots the data on the first subplot. Again, let's take a closer look at the parameters used in this line.

➤ The first two arguments, `months` and `revenue`, provide the x-data and y-data, respectively.

➤ The `marker="o"` parameter instructs Matplotlib to place a solid circle marker at each data point on the line.

➤ The `color="green"` parameter sets the color of the line itself.

You then add context to this first plot using `ax1.set_title()`, `ax1.set_xlabel()`, and `ax1.set_ylabel()` to provide a clear title and axis labels. The exact same process is repeated for the second axes, `ax2`, but this time the expenses data is passed to its `plot()` method and the color is set to red.

Finally, `plt.tight_layout()` is called to automatically adjust the padding between the plots to prevent their titles and labels from overlapping, and `plt.show()` displays the fully rendered figure. On a single "page," you get a clear picture of financial performance from multiple angles.

Customizing Visualizations to Enhance Understanding

Once you have your axes, you can enrich the visualization with elements. Using the data from the previous example, you can add features to your plots like line markers, axis labels, and titles. Listing 3-2 shows multiple data series being combined into a single chart to highlight relationships and comparisons. It also shows how to combine multiple data series in a single chart to highlight relationships and comparisons. Combine the code of Listing 3-1 and Listing 3-2 to create a line plot.

LISTING 3-2: COMBINING ELEMENTS IN CHARTS

```
fig, ax = plt.subplots()

ax.plot(months, revenue, marker="o", label="Revenue")
ax.plot(months, expenses, marker="s", label="Expenses")

ax.set_title("Revenue vs. Expenses Over Time")
ax.set_xlabel("Month")
ax.set_ylabel("Amount ($)")
ax.grid(True, linestyle="--", alpha=0.6)
ax.legend()
```

The code starts by creating a figure and axes with `fig, ax = plt.subplots()`, establishing the canvas and plotting area. On that axes, two line plots are drawn: the first shows monthly revenue in a line with circular markers, while the second shows monthly expenses using square markers. By placing both series on the same chart, it becomes easy to see how revenue consistently stays above expenses and how both measures trend upward over time. To improve readability, the chart is enhanced with several elements: a title, axis labels for months and dollar amounts, dashed gridlines that make it easier to trace values across the plot, and a legend that distinguishes revenue from expenses. Together, these features turn a basic plot into a professional visualization that communicates clearly.

Plotting Options

Once you've selected the right type of plot for your data, your next step is to make it clear, readable, and professional. A "naked" chart without labels is often meaningless. Matplotlib gives you simple, powerful functions to add these crucial context-providing elements. Customizing your plot is how you guide your audience's eye and ensure your data's story is understood correctly. These first few tweaks are the most important, turning raw output into a finished visual. The most critical additions are the contextual labels and titles, which are introduced in Table 3-3.

TABLE 3-3: Basic Chart Customization Options

OPTION	PARAMETERS	DESCRIPTION
Plot title	`ax.set_title()`	Adds a main title to the top of the individual subplot (axes).
X-axis label	`ax.set_xlabel()`	Sets the descriptive label for the horizontal (X) axis.
Y-axis label	`ax.set_ylabel()`	Sets the descriptive label for the vertical (Y) axis.
Legend	`ax.legend()`	Adds a key (legend) to the plot, which is essential if you have multiple datasets (e.g., several lines) on the same chart.

TABLE 3-4: Basic Chart Customization Elements

OPTION/CHART ELEMENT	PARAMETERS	DESCRIPTION
Color	`color=`	A common parameter in most plot functions (like `ax.plot(...,` `color='red')`) that changes the color of the plot's elements.
Linestyle	`linestyle=`	A parameter used in line plots to change the line's style (e.g., solid, dashed, or dotted).
Marker	`marker=`	A parameter used in line and scatterplots to change the style of the data points themselves (e.g., circles, squares, or x).
Figure size	`plt.` `subplots(figsize=(...))`	Sets the overall width and height of the entire figure (the "canvas" holding your plots) when you first create it.

After adding clear chart labels, you can further refine your plot's appearance by styling the visual elements themselves, such as their color, shape, and size. These options, shown in Table 3-4, help distinguish different data series and improve the overall aesthetic of the figure.

This selection only scratches the surface but mastering the eight options shown in Tables 3-3 and 3-4 will dramatically improve the quality and clarity of your visualizations. Think of these as the fundamental grammar of chart design. Later chapters move beyond these basics to explore deeper customizations, such as adjusting axis limits, changing tick marks, adding annotations, and applying advanced styles. For now, simply adding a title and labels is the single best thing you can do to make your plots effective.

CREATING EFFECTIVE VISUALS TO COMMUNICATE BUSINESS DATA

Data visualization is not just about making charts; it is about communicating a message clearly and persuasively. In business contexts, a well-designed chart can highlight opportunities, expose risks, and support decisions that affect millions of dollars. A poorly chosen or poorly designed chart, on the other hand, can obscure the truth, confuse stakeholders, or even lead to the wrong conclusion.

It is important to choose the right chart for a given message. This includes determining the principles of effective design, and practical steps for building visuals that resonate.

The first step in creating an effective visualization is to understand the type of data you are working with. Business data often falls into one of three categories:

- Time-series data

- Cross-sectional data

- Relational data

Each type lends itself to certain chart types, and recognizing this relationship ensures your visuals match the story you want to tell. Using the wrong chart can mislead or distract, while using the right one makes the data's meaning obvious at a glance.

Time-series Data and Line Charts

Time-series data tracks values across a sequence of time periods such as days, months, quarters, or years. Because the main question with this type of data is usually about trends and changes over time, line charts are the most effective choice. They are best used for showing measures like revenue growth, website traffic, stock prices, or sales over months. The strength of a line chart is its ability to clearly show direction and continuity, allowing audiences to easily spot upward or downward movement, seasonal cycles, or volatility. Its limitation is that it is less useful when the goal is to compare multiple categories at a single point in time.

The previous examples, such as Figures 3-1 and 3-2, show line charts. You can refer to those examples for how to create and customize line charts.

Cross-sectional Data and Bar or Pie Charts

Cross-sectional data, on the other hand, represents a snapshot across different categories at one point in time. Examples include revenue by product line in a given quarter or customer counts across regions. In this case, the focus is usually on comparison or proportion, which makes bar charts or pie charts the most effective tools.

Bar charts work best for comparing categories such as sales by region or expenses by department. Their strength is that differences stand out clearly and are easy to interpret, although they can become cluttered when too many categories are included.

Pie charts, by contrast, are designed to show proportions of a whole, such as budget allocation or market share. They are intuitive at a glance when there are only a few slices, but they quickly become hard to read as categories increase and are generally less precise than bar charts. The code in Listing 3-3 is an example of a bar chart, and its output is shown in Figure 3-3.

LISTING 3-3: A BAR CHART EXAMPLE

```
import matplotlib.pyplot as plt

regions = ["North", "South", "East", "West"]
q2_sales = [180000, 150000, 210000, 160000]
```

```
fig, ax = plt.subplots()
ax.bar(regions, q2_sales)
ax.set_title("Q2 Sales by Region")
ax.set_xlabel("Region")
ax.set_ylabel("Sales ($)")
plt.show()
```

The adjustment in code from a line to a bar chart is simple, using `.bar()` rather than `.plot()`. Here the data is cross-sectional: one quarter, multiple regions. A bar chart emphasizes differences across categories, which is exactly the decision task—who's leading, who's lagging, and by how much. The uniform baseline and bar lengths make comparisons effortless, avoiding the precision problems of pies when categories multiply.

Relational Data and Scatterplots

Finally, relational data examines how two variables interact, such as the relationship between advertising spend and sales or between employee training hours and productivity. Scatterplots are particularly well suited for this type of data because they reveal patterns, clusters, and outliers that may indicate correlation or even causation. They are best for helping analysts and decision-makers see whether two metrics move together, identifying unusual values, and spotting natural groupings. Their strength is the ability to make relationships visible, including nonlinear ones, but they require some statistical literacy to interpret and may not feel intuitive to executives unfamiliar with scatterplots. The result of the code in Listing 3-4 is shown in Figure 3-4.

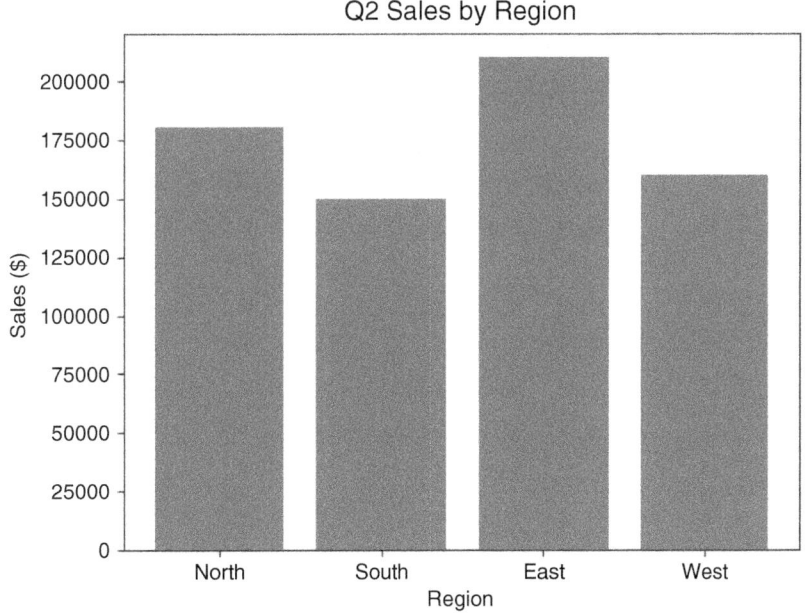

FIGURE 3-3: A bar chart created by Matplotlib.

LISTING 3-4: A SCATTERPLOT CHART

```
import matplotlib.pyplot as plt

ad_spend =  [20, 25, 30, 35, 38, 42, 45, 50]    # in $K
monthly_sales = [210, 230, 260, 280, 295, 320, 330, 360]  # in $K

fig, ax = plt.subplots()
ax.scatter(ad_spend, monthly_sales)
ax.set_title("Advertising Spend vs. Monthly Sales")
ax.set_xlabel("Advertising Spend ($K)")
ax.set_ylabel("Sales ($K)")
ax.grid(True, linestyle="--", alpha=0.6)
plt.show()
```

Similar to the previous example, creating a scatterplot requires using the .scatter() function of Matplotlib. This code answers the question, "How does one variable move with another?" The scatterplot reveals the pattern: as advertising spend increases, sales tend to rise. Clusters and outliers become visible, helping you assess correlation and potential diminishing returns. Unlike a line or bar, the scatter focuses on the relationship rather than a time trend or category comparison.

When the dataset is indexed by time, use a line to communicate direction and tempo. When the dataset is a single snapshot across categories, use a bar for comparisons or a pie for simple proportions. When the goal is to study how two variables move together, use a scatter to expose relationships, clusters, and outliers. Matching chart to data type ensures that the visual aligns with the question and keeps your audience focused on the decision-critical message.

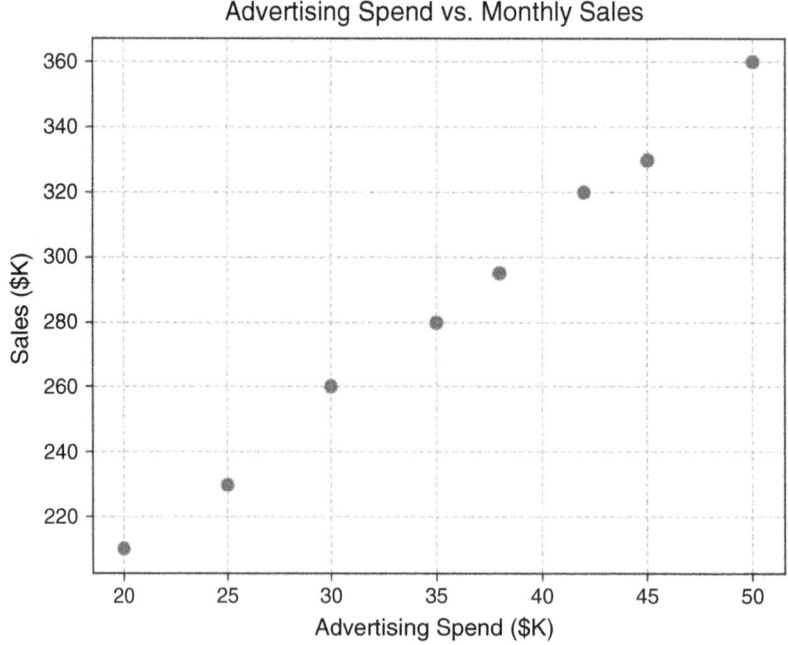

FIGURE 3-4: A scatterplot produced by Matplotlib.

Other Charts You Can Create

When it comes to creating visuals, Matplotlib has so many different options to explore, and choosing the right one is the first step to telling a clear story with your data. Whether you need to show a trend, compare categories, or understand a distribution, there's a specific function designed for the task. A later chapter dives deeper into the customization options. Table 3-5 lays out just a few of the different plot types available, including the ones discussed here.

TABLE 3-5: Matplotlib Chart Types

PLOT TYPE	MATPLOTLIB FUNCTION	DESCRIPTION
Line plot	plt.plot()	Shows trends over time or another continuous sequence (e.g., a stock's price over one year).
Scatterplot	plt.scatter()	Shows the relationship between two different numerical variables (e.g., plotting a person's height vs. their weight).
Bar chart	plt.bar()	Compares a numerical value across different distinct categories (e.g., total sales for different store locations).
Horizontal bar chart	plt.barh()	Same as a bar chart, but better when category names are long (e.g., movie ticket sales by movie title).
Histogram	plt.hist()	Shows the distribution (frequency) of a single numerical variable (e.g., showing how many students scored in the 70s, 80s, 90s, etc., on a test).
Boxplot	plt.boxplot()	Visualizes the summary (median, quartiles, outliers) of one or more datasets (e.g., comparing salary ranges across different job departments).
Pie chart	plt.pie()	Shows the proportions (percentages) of categories that make up a whole (e.g., what percentage of a budget goes to rent, food, and transport).
Heat map	plt.imshow()	Visualizes 2D data (a matrix) using color to show value (e.g., a correlation matrix showing how strongly different variables are related).

As you can see, each function is specialized for a different kind of data story. Don't worry about memorizing every single one right now; you'll quickly build an intuition for which plot to use as you work with different datasets. The key takeaway is that these plots are just the building blocks. The real power of Matplotlib, which is explored in the upcoming chapter on customization, comes from your ability to take these basic charts and refine every single element, from colors and labels to sizes and styles, or even combine multiple plot types into one sophisticated figure.

VISUALIZING TRENDS AND PATTERNS FOR BUSINESS INSIGHTS

Now that you are more familiar with Matplotlib, you can begin to look at more practical and advanced applications. A well-chosen chart can bring out seasonality, uncover anomalies, clarify category comparisons, or highlight cumulative effects that drive strategic decisions. This section explores common approaches to spotting these patterns and provides practical examples of how to implement them in Python.

Highlighting Seasonality and Long-term Growth

When you're working with time-series data, such as monthly revenue, customer sign-ups, or stock prices, line charts are invaluable for showing both short-term fluctuations and long-term growth. Seasonality is especially important in industries like retail, travel, and consumer goods, where demand naturally rises and falls across the year.

For example, plotting monthly revenue for three consecutive years, as in Listing 3-5, can reveal holiday spikes in December or summer slowdowns. A line chart makes these repeating cycles obvious, while also showing whether the overall trend is upward, downward, or flat.

LISTING 3-5: PLOTTING MONTHLY REVENUE

```python
import matplotlib.pyplot as plt

# Sample monthly revenue data for 3 years
months = ["Jan", "Feb", "Mar", "Apr", "May", "Jun",
          "Jul", "Aug", "Sep", "Oct", "Nov", "Dec"]

revenue_2020 = [40000, 42000, 43000, 45000, 44000, 46000,
                47000, 48000, 49000, 51000, 60000, 75000]

revenue_2021 = [42000, 43000, 44000, 46000, 45000, 47000,
                49000, 50000, 52000, 54000, 62000, 77000]

revenue_2022 = [45000, 46000, 47000, 49000, 48000, 50000,
                52000, 53000, 55000, 57000, 65000, 80000]

# Create the figure and axis
fig, ax = plt.subplots(figsize=(9, 5))
```

```
# Plot each year's data
ax.plot(months, revenue_2020, marker="o", label="2020")
ax.plot(months, revenue_2021, marker="s", label="2021")
ax.plot(months, revenue_2022, marker="^", label="2022")

# Add titles and labels
ax.set_title("Monthly Revenue Across Three Years")
ax.set_xlabel("Month")
ax.set_ylabel("Revenue ($)")
ax.grid(True, linestyle="--", alpha=0.6)
ax.legend()

plt.tight_layout()
plt.show()
```

The results of running the code in Listing 3-5 are the line chart shown in Figure 3-5. This code creates a line chart that shows monthly revenue across three years. It begins by listing the months and then defining revenue numbers for 2020, 2021, and 2022. The `plt.subplots()` command sets up the chart, and `ax.plot()` is used three times to draw one line for each year, with different markers so they are easy to tell apart. Titles and axis labels are added to explain what the chart shows, and `ax.grid()` creates faint gridlines to make the numbers easier to read. Finally, `ax.legend()` adds a small box that labels each line. When you run the code, `plt.show()` displays the finished chart.

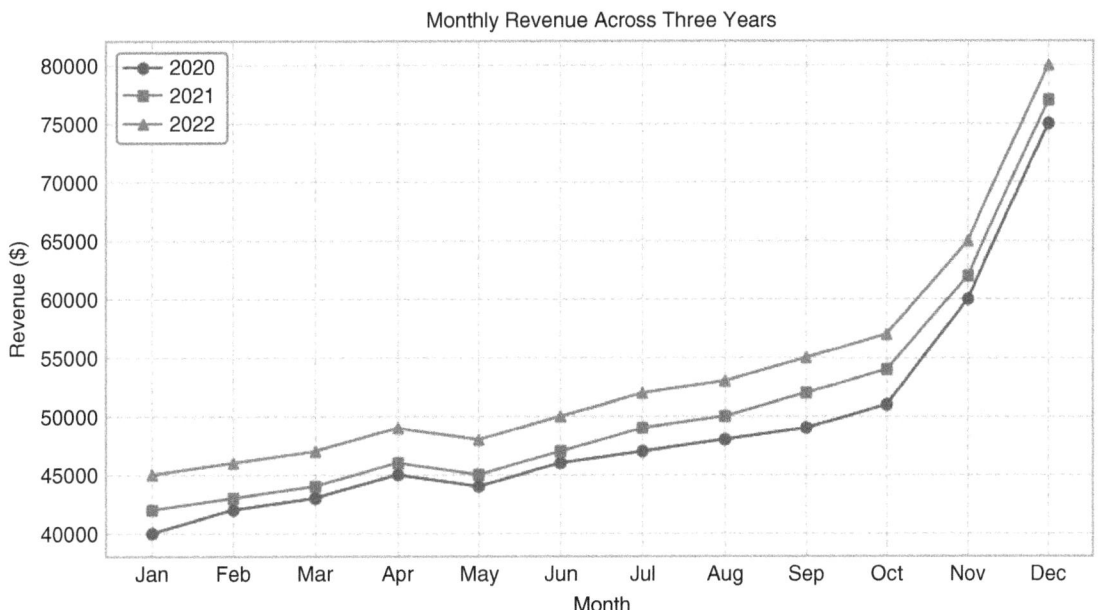

FIGURE 3-5: A line chart showing seasonal revenues increasing in December.

Comparing Categories and Segments

It is often not enough to simply track overall performance. Decision-makers want to know how different categories or segments contribute to results and how they compare across time periods or markets. Bar charts are the natural choice for this type of analysis, but when more detail is required, stacked bar charts and grouped bar charts provide richer views of the data.

A *stacked bar chart* shows both the total and the contribution of each category within that total. This makes it ideal for understanding composition. For example, if a company's quarterly revenue comes from four product lines, stacking them in a single bar lets managers quickly see the overall revenue per quarter while also understanding which products are driving growth. If one product line consistently takes up most of the bar, it signals dependence on that product.

In contrast, a *grouped bar chart* places categories side by side, making it easier to directly compare their magnitudes. Using the same example, a grouped bar chart would plot the four product lines next to each other within each quarter. This layout emphasizes differences between categories, helping managers identify which products are outperforming others in any given quarter.

The key distinction is simple: stacked bars highlight contributions to a whole, while grouped bars highlight comparisons between categories. Both perspectives are useful, but they tell different stories.

Listing 3-6 shows how the same dataset can be visualized with both a stacked bar chart and a grouped bar chart. This listing is an example of tracking quarterly revenue by product line.

LISTING 3-6: USING STACKED AND GROUPED BAR CHARTS

```python
import matplotlib.pyplot as plt
import numpy as np

quarters = ["Q1", "Q2", "Q3", "Q4"]
electronics = [20000, 22000, 25000, 27000]
clothing = [15000, 16000, 18000, 19000]
groceries = [30000, 31000, 32000, 33000]

# Stacked Bar Chart
fig, ax = plt.subplots(figsize=(8, 5))

ax.bar(quarters, electronics, label="Electronics")
ax.bar(quarters, clothing, bottom=electronics, label="Clothing")
# stacking groceries on top of electronics+clothing
bottom_stack = np.array(electronics) + np.array(clothing)
ax.bar(quarters, groceries, bottom=bottom_stack, label="Groceries")

ax.set_title("Quarterly Revenue by Product Line (Stacked)")
ax.set_ylabel("Revenue ($)")
ax.legend()
plt.show()

# Grouped Bar Chart
x = np.arange(len(quarters))  # numeric positions for quarters
width = 0.25  # width of each bar
```

```
fig, ax = plt.subplots(figsize=(8, 5))

ax.bar(x - width, electronics, width, label="Electronics")
ax.bar(x, clothing, width, label="Clothing")
ax.bar(x + width, groceries, width, label="Groceries")

ax.set_xticks(x)
ax.set_xticklabels(quarters)
ax.set_title("Quarterly Revenue by Product Line (Grouped)")
ax.set_ylabel("Revenue ($)")
ax.legend()
plt.show()
```

In the stacked bar chart generated by this listing (shown in Figure 3-6), each quarter is represented by a single bar, but the bar is divided into segments for Electronics, Clothing, and Groceries. This makes it easy to see both the total revenue per quarter and the composition of that revenue. For instance, if Groceries consistently takes up the largest portion of each bar, it signals that the company relies heavily on grocery sales to sustain its revenue base. The stacked format also reveals whether the mix of products is changing, for example, if Electronics begins to grow faster and occupies a larger share of the bar over time.

In the grouped bar chart (shown in Figure 3-7), the three product lines are placed side by side within each quarter. This makes direct comparison between categories more straightforward. If Electronics

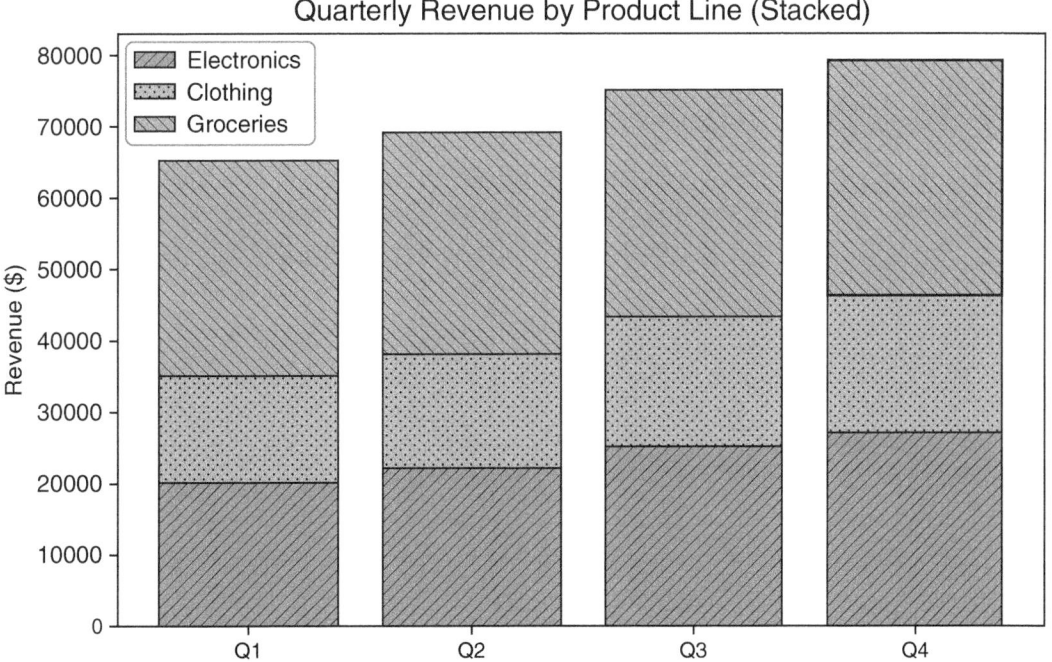

FIGURE 3-6: A stacked bar chart produced by Matplotlib.

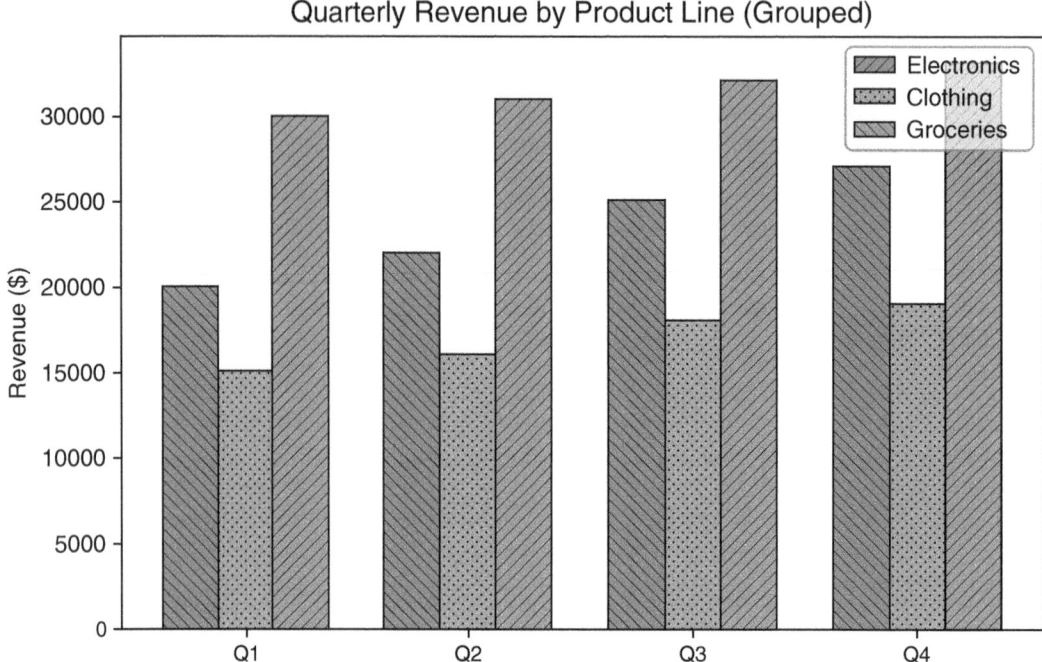

FIGURE 3-7: A grouped bar chart produced by Matplotlib.

and Clothing are plotted next to each other, managers can immediately see that Electronics outperforms Clothing in every quarter, while Groceries dominate overall. Grouped bars are especially useful when the main question is, "Which category is bigger? " rather than "What proportion does each category contribute to the whole?"

Together, these two visualization styles provide complementary perspectives. The stacked chart emphasizes the total picture and contributions, which is important when managing overall revenue or market share. The grouped chart emphasizes head-to-head comparisons, which helps identify winners and laggards. By applying both, analysts can provide management with a more complete understanding of category performance.

Visualizing Cumulative Effects

Outcomes often build over time. Investors track cumulative returns to see how a stock has performed from the beginning of an investment, while managers may track cumulative sales to see whether annual goals are on pace to be met.

Cumulative plots are powerful because they emphasize momentum and progress. A chart of cumulative sales, for instance, can show whether the company is ahead or behind target as the year unfolds. Unlike monthly snapshots, cumulative curves provide a running total, reinforcing the importance of sustained performance.

As a practical example, let's visualize a company's monthly revenue for a single year. Listing 3-7 provides a hands-on example of visualizing this time series.

LISTING 3-7: VISUALIZING A SERIES WITH PYTHON

```python
import matplotlib.pyplot as plt
import numpy as np

months = ["Jan", "Feb", "Mar", "Apr", "May", "Jun",
          "Jul", "Aug", "Sep", "Oct", "Nov", "Dec"]
revenue = [42000, 43000, 45000, 48000, 47000, 49000,
           52000, 51000, 53000, 55000, 60000, 75000]

# Calculate cumulative revenue
cumulative_revenue = np.cumsum(revenue)

fig, ax = plt.subplots()

# Plot cumulative revenue instead of monthly
ax.plot(months, cumulative_revenue, marker="o", color="blue")
ax.set_title("Cumulative Revenue (2022)")
ax.set_xlabel("Month")
ax.set_ylabel("Cumulative Revenue ($)")
ax.grid(True, linestyle="--", alpha=0.6)
plt.show()
```

The results of running this code are shown in Figure 3-8. This example demonstrates how a simple line chart can be used to track how revenue accumulates over the course of a year. The code begins by defining the months of the year along with the monthly revenue values, then applies the `cumsum` function from NumPy to calculate a running total. Using `fig, ax = plt.subplots()`, the chart is set up, and the cumulative revenue is plotted against the months with a blue line and circular markers to highlight each point. Titles and axis labels clearly explain what the chart represents, while gridlines make it easier to follow the progression across months.

The resulting chart tells a different kind of business story: instead of showing individual monthly fluctuations, it emphasizes the total revenue that builds month after month. By December, the cumulative line makes the dramatic end-of-year surge even clearer, showing how much of the company's annual results depend on holiday demand. This view is especially useful for managers and executives who want to measure progress toward goals and understand how each month contributes to the yearly total.

Smoothing Trends with Rolling Averages

While line charts are excellent for showing raw performance over time, real-world business data often fluctuates from month to month due to seasonality, promotions, or random noise. These ups and downs can make it difficult to see the underlying trend. One-way analysts address this is by using a *rolling average* (also called a *moving average*), which smooths short-term volatility by averaging performance across a fixed window of time.

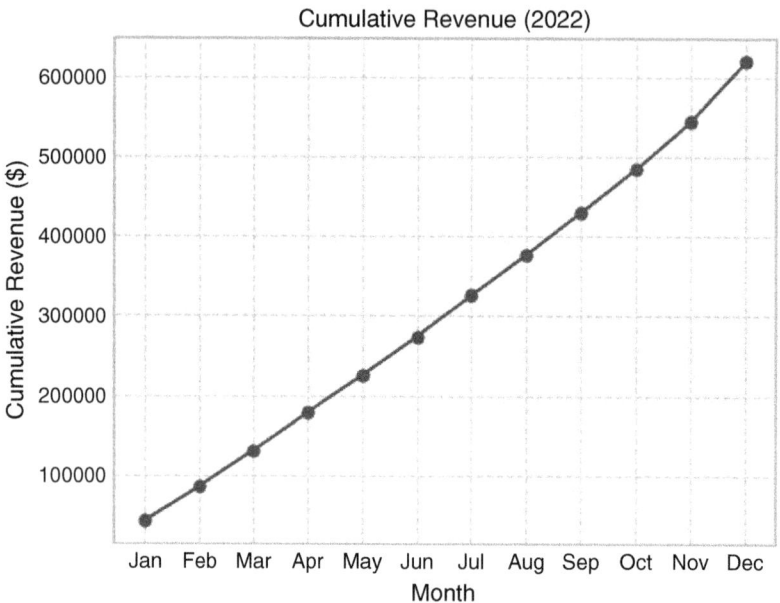

FIGURE 3-8: A cumulative line chart produced by Matplotlib.

For example, consider a retailer analyzing monthly revenue. Instead of focusing only on the raw month-to-month values, they might calculate a three-month rolling average, which takes each month's revenue and averages it with the two preceding months. This approach, as shown in Listing 3-8, filters out temporary spikes or dips and highlights the longer-term trajectory.

LISTING 3-8: SMOOTHING TRENDS

```python
import matplotlib.pyplot as plt
import pandas as pd

# Monthly revenue data
months = ["Jan", "Feb", "Mar", "Apr", "May", "Jun", "Jul",
          "Aug", "Sep", "Oct", "Nov", "Dec"]
revenue = [42000, 43000, 45000, 48000, 47000, 49000,
           52000, 51000, 53000, 55000, 60000, 75000]

# Put into a pandas Series for rolling calculation
revenue_series = pd.Series(revenue, index=months)

# 3-month rolling average
rolling_avg = revenue_series.rolling(window=3).mean()

fig, ax = plt.subplots(figsize=(8, 5))

# Plot raw revenue
ax.plot(months, revenue, marker="o", color="blue", label="Monthly Revenue")
```

```
# Plot rolling average
ax.plot(months, rolling_avg, marker="s", color="orange", line-
style="--", label="3-Month Rolling Avg")

ax.set_title("Monthly Revenue with Rolling Average (2022)")
ax.set_xlabel("Month")
ax.set_ylabel("Revenue ($)")
ax.grid(True, linestyle="--", alpha=0.6)
ax.legend()

plt.show()
```

This chart plots two series—as shown in Figure 3-9—the raw monthly revenue (the solid line) and a smoothed three-month rolling average (the dashed line). The raw line shows natural ups and downs (like the dip in May and the spike in December), but the rolling average reveals a smoother trajectory that highlights the overall growth trend. A sales manager, for instance, might use the rolling average to report consistent progress to leadership, while also using the raw data to plan around short-term demand cycles.

Line Charts with Confidence Intervals Using Seaborn

In real business datasets, such as dozens of stores reporting revenue each month or hundreds of SKUs contributing to sales, it is rarely enough to show just one raw series. A manager or executive often wants to know not only what the average performance looks like, but also how much uncertainty there is

FIGURE 3-9: A line chart with a three-month rolling average.

around that average. This is where confidence intervals come in, as shown in Listing 3-9. Instead of presenting a single line, a confidence interval wraps the line with a shaded band, giving the audience a sense of how stable the trend is. If the band is narrow, it suggests consistency across stores or products; if the band is wide, it signals variability and risk. The results of running this code are found in Figure 3-10.

LISTING 3-9: USING CONFIDENCE INTERVALS

```
import numpy as np
import pandas as pd
import matplotlib.pyplot as plt
import seaborn as sns

# 1) Simulate tidy business data: multiple stores reporting monthly revenue
rng = np.random.default_rng(7)
months = pd.Index(
    ["Jan","Feb","Mar","Apr","May","Jun","Jul","Aug","Sep","Oct","Nov","Dec"],
    name="month"
)
n_stores = 40

# Base seasonal pattern: gentle climb + December spike
seasonal = np.array([42, 43, 45, 48, 47, 49, 52, 51, 53, 55, 60, 75]) * 1000

# Store-level deviations + random noise
records = []
for store in range(n_stores):
    store_effect = rng.normal(0, 2500)
    noise = rng.normal(0, 2000, size=len(months))
    rev = seasonal + store_effect + noise
    for m, r in zip(months, rev):
        records.append({"store_id": store, "month": m, "revenue": r})

df = pd.DataFrame(records)

# 2) Plot with seaborn: mean revenue per month + 95% confidence interval
sns.set_theme(style="whitegrid")

fig, ax = plt.subplots(figsize=(9, 5))
sns.lineplot(
    data=df,
    x="month",
    y="revenue",
    errorbar=("ci", 95),
    estimator="mean",
    n_boot=1000,
    ax=ax
)

ax.set_title("Average Monthly Revenue with 95% Confidence Interval")
ax.set_xlabel("Month")
ax.set_ylabel("Revenue ($)")
plt.tight_layout()
plt.show()
```

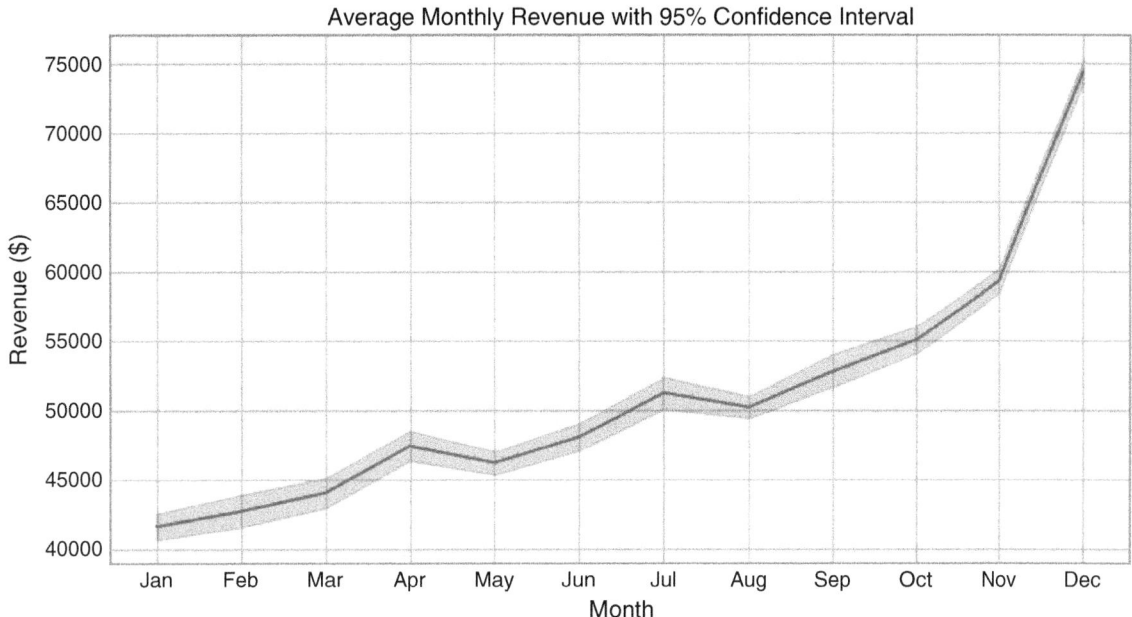

FIGURE 3-10: A line chart with confidence intervals made with Seaborn.

Using the Seaborn package makes the style of visualization shown in Figure 3-10 straightforward. Unlike earlier examples that relied solely on Matplotlib, Seaborn is designed for tidy datasets and can automatically compute averages, bootstrap confidence intervals, and display them in a polished way.

As noted, Seaborn is a separate Python library built on top of Matplotlib. This means it needs to be installed before it can be used. You can typically install it using `pip` from your terminal: `pip install seaborn`.

The primary advantage of Seaborn is its deep integration with pandas DataFrames. Where Matplotlib often requires you to manually pull out columns or data series, Seaborn's plotting functions are designed to work directly with DataFrame columns, letting you specify variable names as strings. This, combined with its statistical power, is what allows it to create complex plots with very little code.

Seaborn's specialty is statistical storytelling with "tidy" data (like pandas DataFrames). Instead of asking you to draw lines and boxes, Seaborn asks, "What's the story you want to tell about your data?" For example:

➤ When you use Matplotlib, you say: "Draw a red line using these X points and these Y points."

➤ When you use Seaborn, you say: "Here is my entire dataset. I want to see the relationship between the Price column and the Time column, and please group everything by the Asset Class column."

In the example, monthly revenue is simulated across 40 stores. Seaborn's `lineplot` function calculates the mean revenue for each month and surrounds it with a shaded 95% confidence band. The result is a clean, smooth line that shows the average pattern of revenue, along with a ribbon that reflects how consistent or inconsistent stores are around that average. December's spike is still visible, but now the audience also learns how reliable that surge is across different stores, a nuance that is essential for planning and risk management.

Confidence intervals are not limited to time series. They can also be used when exploring relationships between variables. A scatterplot of advertising spend and sales, for instance, shows individual observations but does not quantify the overall trend. By using Seaborn's `regplot`, you can add a fitted regression line with a confidence band around it. The line suggests the general direction of the relationship—higher advertising spend is associated with higher sales—while the band reveals the uncertainty in that estimate. A narrow band signals a stable relationship, while a wide band warns that sales vary considerably even at similar spending levels.

This type of visualization is more advanced than the earlier line charts, bar charts, and scatterplots because it goes beyond description. Rather than just showing raw values, it introduces estimation and uncertainty, offering the audience both the likely pattern and the degree of confidence in that pattern. For executives, this shift is critical. A single line might encourage overconfidence, but a line with a confidence band communicates both opportunity and risk. In decision-making settings, that balance is far more powerful than the illusion of certainty.

Analyzing Relationships and Distributions with *jointplot*

The previous example explored a time-series trend. Another common business task is to understand the relationship between two continuous variables. For example, how does a fund's volatility relate to its annual return? A simple Matplotlib scatterplot can show the data points, but it's a very limited view. It doesn't quantify the trend, nor does it show the distribution of each variable on its own. Are most funds low-volatility? Are the returns normally distributed? Answering these questions requires a more advanced visualization.

This is where Seaborn's `jointplot` function excels. As shown in Listing 3-10, `jointplot` is a compound or figure-level function. It automatically creates a figure that combines three plots into one—a central scatterplot to show the relationship, and two histograms (one on the top margin and one on the right margin) to show the individual distributions of the x-axis and y-axis variables.

LISTING 3-10: USING JOINTPLOT TO SHOW RELATIONSHIPS AND DISTRIBUTIONS

```python
import numpy as np
import pandas as pd
import matplotlib.pyplot as plt
import seaborn as sns

# 1) Simulate tidy financial data: volatility vs. return for 150 funds
rng = np.random.default_rng(42)
n_funds = 150

# Create correlated data: higher volatility tends to mean higher returns,
```

```
# but with a lot of variance.
volatility = rng.normal(loc=18, scale=4, size=n_funds)
# Make returns dependent on volatility + random noise
returns = (volatility * 0.5) + rng.normal(loc=2, scale=3, size=n_funds)

# Put into a DataFrame
funds_df = pd.DataFrame({
    "annual_volatility_pct": volatility,
    "annual_return_pct": returns
})

# 2) Plot with seaborn: show relationship AND individual distributions
# This is a "figure-level" plot, so we don't create a plt.subplots() first.
# Seaborn handles the figure creation.
g = sns.jointplot(
    data=funds_df,
    x="annual_volatility_pct",
    y="annual_return_pct",
    kind="reg",
    height=7,
    color="royalblue",

    # --- Customization Dictionaries ---
    joint_kws={
        "scatter_kws": { 's': 40, 'alpha': 0.6 }
    },
    marginal_kws={
        'bins': 20, 'kde': True, 'color': 'gray'
    }
)

# 3) Add titles and labels (syntax is slightly different)
g.set_axis_labels("Annual Volatility (%)", "Annual Return (%)")
g.fig.suptitle("Relationship Between Fund Volatility and Return", y=1.02)

plt.show()
```

This example shows how Seaborn combines statistical power with deep customization. Unlike the `lineplot` function (an "axes-level" function), `jointplot` is a "figure-level" function, meaning it creates and manages its own Matplotlib figure.

Let's break down the customization in Listing 3-10:

> `kind="reg"`: This is the most important parameter. By default, `jointplot` just shows a scatterplot. By setting `kind="reg,"` it instructs Seaborn to automatically run a linear regression on the data and—just like in the `lineplot` example—draw the resulting trend line along with its 95% confidence interval band.

> `height=7`: Since Seaborn is creating the figure, you can control the size (in inches) with this parameter.

> `joint_kws={...}`: This is a powerful customization feature. It's a dictionary of "keyword arguments" that are passed directly to the underlying plot function in the central panel. Since

the `kind="reg"` plot has scatter points, you can pass a `scatter_kws` dictionary inside it to make the points semi-transparent (`alpha=0.6`) and smaller (`s=40`).

➤ `marginal_kws={...}`: Similarly, this dictionary passes arguments to the two marginal histograms. This code tells Seaborn to use 20 bins for the histograms, to set their color to gray, and to overlay a smooth Kernel Density Estimate (KDE) line (`kde=True`) on both.

The results shown in Figure 3-11 tell a complete story, offering far more insight than a simple scatterplot ever could. A risk manager or investor can immediately see the trend from the central regression line, which confirms the expected positive relationship: higher volatility is generally associated with higher returns. But just as quickly, their eye is drawn to the uncertainty. The wide, shaded confidence band reveals that this relationship isn't very precise and that there's a great deal of variance; you can't simply assume volatility guarantees a specific return.

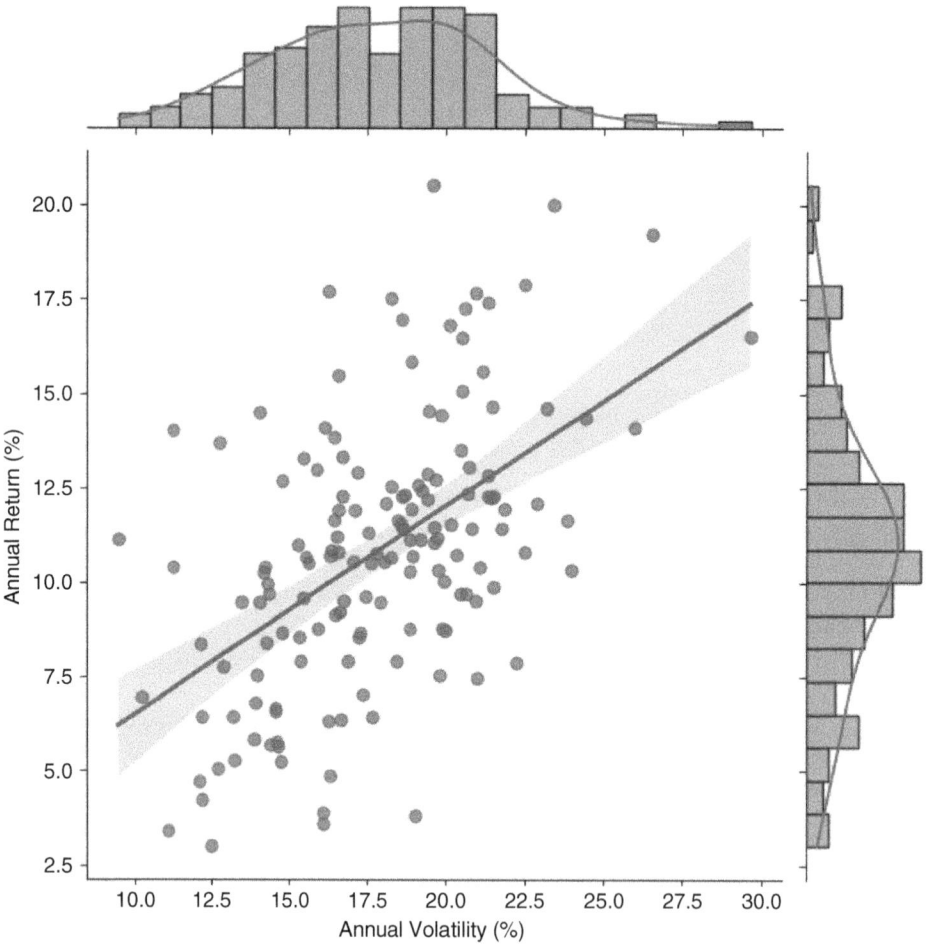

FIGURE 3-11: A joint plot with regression line, confidence interval, and marginal KDEs.

At the same time, the marginal plots provide crucial context that a simple scatterplot omits. The top histogram shows the X-distribution, revealing that most funds cluster in the 15–20% volatility range, with very few existing in the high-risk, high-volatility space. The right histogram shows the Y-distribution, illustrating the spread of returns, which appear centered around 8–10%. In one glance, the analyst gets the trend, the risk (uncertainty), and the shape of the data. This ability to automatically compute statistics (like the regression line, CI, and KDEs) and combine multiple plot types into one cohesive view, all while offering deep customization via _kws dictionaries, is what makes Seaborn an indispensable tool for rapid data exploration.

SUMMARY

In this chapter, you learned how visualization turns numbers into stories. You first learned about the basics of Matplotlib, discovering how figures and axes create the canvas for plotting. You saw how to produce simple line and bar charts, the fundamental tools for any analysis.

From there, the chapter explored how different chart types map to specific financial data: line charts for tracking time-series data like stock prices, bar charts for comparing cross-sectional data like portfolio sector weights, and scatterplots for understanding relational data, such as the correlation between two different assets. I stressed that clarity is always more important than decoration; choosing the right chart is essential to avoid misinterpreting a market signal.

As the chapter progressed, the discussion moved beyond simple plots to uncover deeper quantitative insights. Line charts reveal market trends and seasonality, while scatterplots and boxplots help identify outliers like volatility spikes or anomalous trades. Stacked and grouped bars were used to compare the performance of different strategies or asset classes.

I also introduced plots crucial to finance, like cumulative return (P&L) curves to track strategy performance, rolling averages to smooth price volatility, and year-over-year comparisons to contextualize quarterly earnings. Finally, the chapter stepped into more advanced territory with Seaborn, adding confidence intervals to these plots, showing not just a central trend, but the statistical uncertainty and risk surrounding it.

Taken together, this chapter highlighted that visualization isn't about making data look pretty; it's about making data actionable. The right chart helps a quant or portfolio manager see risks before they materialize, highlight alpha opportunities that raw numbers would bury, and build narratives that guide investment strategy. Whether it's a simple line chart for a pitch or an advanced plot for forecasting risk, the principles are the same—choose the chart that fits the data, design for clarity, and use the design to communicate insights that support smarter, data-driven financial decisions.

CONTINUE YOUR LEARNING

The following are some resources you may find useful:

> ➤ **Matplotlib cheat sheets:** These cheat sheets help you create and customize visualizations. You can find them at: `https://matplotlib.org/cheatsheets/`

➤ **Matplotlib gallery:** Numerous examples of charts from time-series data to choropleth maps. You can find the gallery of charts at: `https://matplotlib.org/stable/gallery/index.html`

➤ **HvPlot gallery:** Examples of figures and charts made using the HvPlot library. You can find these examples at: `https://hvplot.holoviz.org/en/docs/latest/gallery/index.html`

➤ **Seaborn example gallery:** More examples of figures created using the Seaborn library. You can find these examples at: `https://seaborn.pydata.org/examples/index.html`

PART 2
Applying the Math

➤**Chapter 4:** Linear Algebra for Business and Finance

➤**Chapter 5:** Calculus for Business Problem Solving

➤**Chapter 6:** Optimization Techniques for Business Strategy

➤**Chapter 7:** Probability and Statistics for Business Analytics

➤**Chapter 8:** Applied Business Problems with Math and Python

Linear Algebra for Business and Finance

Linear algebra may sound like a topic reserved for an advanced math class, but it's the engine behind many of the most powerful algorithms and solutions to business problems. This chapter breaks down this essential subject by introducing its core building blocks: vectors and matrices. The chapter then moves beyond abstract theory to show how these concepts provide a practical language for organizing information. You'll see how a simple vector can represent anything from regional sales to an investment portfolio, while a matrix can capture the complex dynamics of customer transitions or the interconnected risks within the stock market. The goal is to translate these mathematical structures into a concrete framework for making smarter, data-driven decisions.

Of course, understanding the concepts is just the first step; the real magic happens when you apply them. You'll see how to use Python's NumPy library to bring these ideas to life, starting with basic operations and building toward a complete, real-world financial analysis. This journey will take you from calculating revenue with a simple dot product to modeling and comparing entire investment strategies. Finally, this chapter explores one of linear algebra's most profound concepts, *eigenvectors*, to uncover the hidden, long-term trends in your data.

WORKING WITH VECTORS AND MATRICES

At its heart, linear algebra is a powerful language that helps people solve and see many common challenges in a new light. Think of it as a toolkit for organizing and working with numeric information. This chapter starts with two of the most important tools in that kit: vectors and matrices.

A *vector* is perfect for representing simple lists of related numbers. Imagine you want to track your company's sales in different cities, the cash flow you expect each month for the next year, or how you've divided up your investment portfolio. A vector can hold all that information in a clean, organized way.

A *matrix*, on the other hand, helps you tackle more complex situations where multiple factors are at play. You can use a matrix to map out how different parts of the economy rely on each other, to predict how customers might switch between different products, or to understand how the returns of different stocks move together.

While vectors and matrices might seem like two completely different things at first, you will soon see that they're closely related. In fact, the best way to think of a matrix is simply as a collection of vectors working together.

Understanding Vectors

As I touched on in a previous chapter, a vector is just an ordered list of numbers. You'll usually see it written as a column of numbers, but it can also be a row.

Vectors are incredibly useful because they give structure to data. They allow you to group related numbers together so you can work with them as a single unit. This is a lot more efficient than juggling several individual figures.

Some examples of vectors in practice include:

➤   **A revenue vector:** Imagine your company sells five different products. You could create a revenue vector where each entry represents the total sales for one of those products. This gives you a quick snapshot of your entire product line's performance.

➤   **A portfolio weight vector:** If you're an investor, you could use a vector to represent how your money is split between different assets like stocks, bonds, and real estate. Each entry in the vector would be the percentage of your total capital allocated to that asset.

Let's look at a simple sales example. Suppose you have a vector that represents the number of units sold for three different product lines:

$$Q = \begin{bmatrix} 100 \\ 200 \\ 150 \end{bmatrix}$$

The vector (here named Q for quantity) tells you, in a very neat and tidy way, that you sold 100 units of the first product, 200 units of the second, and 150 units of the third.

The real power of vectors comes from the fact that you can perform mathematical operations on the entire list of numbers at once. Instead of having to work with each product's sales figures individually, you can use the tools of linear algebra to analyze and manipulate all of the sales data in one go. This saves time, reduces the chance of errors, and allows you to uncover insights that would be difficult to see otherwise.

Understanding Matrix

You've seen how a vector can neatly organize sales figures for the three product lines for a single year. But what happens when the data gains another layer of complexity? For instance, imagine you want to track sales not just for one year, but for several.

You could create a separate vector for 2024, another for 2025, and so on. But this quickly becomes a bookkeeping nightmare. Imagine trying to compare sales across 50 years—you'd be juggling 50 different variables and it would become a mess, quickly. So, what about the other idea: just dumping all the data into one "very, very long" vector? This is even worse. You'd lose all the structure. How would you know where 2024's data ends and 2025's begins? You'd have no easy way to ask, "What was the data for 2026?" or "Show me the sales for every year." You'd just have a meaningless sea of numbers.

This is where the matrix comes in. Instead of juggling multiple vectors, you can combine them into a single, more powerful structure, a matrix. It combines all the data into a single structure, so you're not juggling dozens of variables, and it preserves the grouping of each vector (e.g., as a row), so you instantly know that "row 0" is 2024, "row 1" is 2025, and so on.

Think of a matrix as a grid or a spreadsheet, a two-dimensional arrangement of numbers with both rows and columns. Following the sales example, you can take the sales vectors for 2024 and 2025 and stack them side-by-side.

$$M = \begin{bmatrix} 100 & 120 \\ 200 & 210 \\ 150 & 160 \end{bmatrix}$$

Here, each column is a vector of sales for a particular year. The first column corresponds to 2024, the second to 2025. This simple example reveals the true nature of a matrix: it's just a collection of related vectors, organized in a way that provides much richer view of the data. Now you can analyze trends over time and across products all at once.

The real magic of matrices is their ability to represent and analyze complex, multi-faceted relationships. Here are a few examples:

➤ **Shipping and logistics:** A cost matrix can map shipping costs from various warehouses (the rows) to every retail store (the columns). Each number in the matrix shows the exact cost for one specific route, helping businesses instantly find the cheapest options and optimize their supply chain.

➤ **Customer behavior:** A transition matrix models how customers move between different states (e.g., Active, Inactive, Churned) over time. By showing the probability of a customer moving from a starting state to an ending state, it helps companies predict future revenue and design effective retention campaigns.

➤ **Financial risk management:** In finance, a covariance matrix is essential for managing portfolio risk. It captures how different investments, like stocks and bonds, tend to move in relation to each other. This insight allows investors to build diversified portfolios where the risk of one asset is offset by another, aiming for more stable returns.

OPERATIONS WITH VECTORS AND MATRICES

The best way to truly appreciate the power of vectors and matrices is to see them in action. This example uses Python's most popular library for numerical computing, NumPy, which is specifically designed to make working with these structures fast and intuitive.

Let's start by setting up a simple scenario. Imagine your company sells three core products, and you have the sales data for the last two years:

```
import numpy as np

# Sales quantities (vectors)
q_2024 = np.array([100, 200, 150])
q_2025 = np.array([120, 210, 160])

# Prices per unit
prices = np.array([5, 10, 15])
```

Here, q_2024 and q_2025 are vectors holding the number of units sold for each of the three products in those years. The prices vector holds the corresponding price for each product. In NumPy, these are just simple arrays.

One of the most straightforward operations is also one of the most useful: comparing two vectors. As can be seen by using the following code snippet, if you subtract the 2024 sales vector from the 2025 vector, you can see the growth (or decline) for each product line, element by element.

```
import numpy as np
# Change in sales from 2024 to 2025
diff = q_2025 - q_2024
print(diff)
```

The result of running this code snippet is [20 10 10].

This tells you that between 2024 and 2025, sales grew by 20 units for product 1, 10 units for product 2, and 10 units for product 3. The math directly mirrors the business question you wanted to answer.

Using this same data, you can perform other operations on your vectors and matrices. These include fundamental calculations like scalar multiplication and the dot product, measuring vector lengths with norms, and performing advanced matrix operations such as multiplication and transposition. You can also use slicing to surgically access specific rows or columns of your data for targeted analysis.

Scalar Multiplication

What if you wanted to forecast a 10% increase in sales for next year across all product lines? This is where scalar multiplication comes in. A *scalar* is just a fancy word for a single number. Multiplying a vector by a scalar applies that number to every element in the vector. The following code applies the 10% to the 2024 numbers:

```
import numpy as np
# Scale 2024 sales by 10%
q_2024 = np.array([100, 200, 150])
scaled_q = 1.1 * q_2024
print(scaled_q)
```

The result, `[110. 220. 165.]`, shows the effect of forecasting a 10% increase in all product sales. Every entry in the vector grows proportionally.

The Dot Product

The dot product is one of the most important operations in all of linear algebra, especially in business. It works by multiplying the corresponding elements of two vectors and then summing the results.

$$p \cdot q = p_1 q_1 + \cdots + p_n q_n$$

In plain language, the dot product measures how much two vectors "line up" with each other. In business, it often means combining prices and quantities to compute revenue, or portfolio weights and asset returns to compute overall return. In Python, the dot product is handled by NumPy's `np.dot` function. In the following code snippet, you can compute total revenue by finding the dot product between the `prices` and the `q_2024` vectors.

```
import numpy as np
# Total revenue in 2024
q_2024 = np.array([100, 200, 150])
prices = np.array([5, 10, 15])
revenue_2024 = np.dot(prices, q_2024)
print(revenue_2024)
```

The dot product executed the calculation $(5 \times 100) + (10 \times 200) + (15 \times 150)$ in a single, optimized step. The result, `4750`, is the total revenue for 2024. This operation is fundamental for everything from calculating portfolio returns to machine learning.

Norms (Vector Lengths)

The *norm* of a vector measures its size or magnitude. Think of it as the distance from the origin to the point represented by the vector. In business, norms can help measure total exposure, intensity of demand, or even compare the "scale" of different operations.

While there are several ways to calculate a vector's "size," the most common and intuitive method is the Euclidean norm, also known as the L2 norm. This is exactly what it sounds like: it calculates the straight-line distance from the origin $(0, 0, 0)$ to the vector's point in space. It's a direct, multi-dimensional application of the Pythagorean theorem. To calculate this in Python, you use the `norm` function, which is located inside NumPy's linear algebra submodule, `linalg`. This is a sub-package within the NumPy library in Python, providing a set of functions for performing linear algebra operations.

Using the simple helper function, you can compute the norm with the following code snippet:

```
import numpy as np
# Calculate the magnitude of the 2024 sales vector
q_2024 = np.array([100, 200, 150])sales_norm_2024 =
np.linalg.norm(q_2024)print(sales_norm_2024)
```

The result of this code is `269.258`. The `np.linalg.norm` function is performing a calculation you're already familiar with from the dot product. It's effectively squaring every element in the vector (`100**2 + 200**2 + 150**2`), summing them, and then taking the square root. This is similar to the Pythagorean theorem. In fact, this is directly related to the dot product: the dot product of a vector with itself (e.g., `np.dot(q_2024, q_2024)`) gives you the squared norm (`72500`). The result from this code snippet is the square root of that number. While this number is more abstract than total revenue, it gives you a single metric to summarize the vector's size, which is incredibly useful for more advanced comparisons and algorithms.

Combining Matrices

So far, you've learned about vectors, but what if you have two years' worth of sales data? You need a simple way to combine these into a single variable to make your analysis easier, allowing you to perform operations on multiple years at once. This is where matrices come in.

To build the sales matrix, you start, as always, with NumPy and the two sales vectors. You need a way to combine them, and the `np.column_stack` function is designed for exactly this. You give this function your list of vectors, `[q_2024, q_2025]`, and it neatly stacks them side-by-side, turning each vector into a vertical column in the new 2D matrix, `Total_Quantity`. Consider the following example, which builds a sales matrix where each column is a year:

```
import numpy as np

# Sales quantities (vectors)
q_2024 = np.array([100, 200, 150])
q_2025 = np.array([120, 210, 160])

Total_Quantity= np.column_stack([q_2024, q_2025])
print(Q)
```

Just like in the previous examples, you start with your toolkit, `numpy`, and the two vectors of sales quantities. You need a way to combine everything into one matrix. This is where `np.column_stack` comes in. You give this function your list of vectors, `[q_2024, q_2025]`, and it neatly stacks them side-by-side, turning each vector into a vertical column in a new 2D matrix. The `np.column_stack` function can work with 1D vectors or 2D matrices. This new, combined table is stored in the variable Q. Finally, if you call `print(Q)`, you see this result:

$$\begin{bmatrix} 100 & 120 \\ 200 & 210 \\ 150 & 160 \end{bmatrix}$$

Now your data is in one place. You can clearly see that the rows represent the three products (at indices 0, 1, and 2), and the columns represent the years (2024 at index 0 and 2025 at index 1). This organized structure is much more powerful. You can now access all of 2024's data by simply grabbing the first column or see the full history for the second product by grabbing the second row.

Slicing Matrices

A matrix is a fantastic way to store a complete dataset, such as the sales performance for all your products across all your regions. But in real-world analysis, you'll rarely work with the entire matrix at once. More often, you'll want to isolate a specific piece of it. For example, what if you want to retrieve only 2024 sales again? This process of grabbing specific parts of your matrix is called *slicing*.

NumPy uses a powerful and intuitive bracket notation, `matrix[row_selector, column_selector]`, to select data. The key thing to remember is that, like all of Python, NumPy is zero-indexed, meaning the first row is at index 0, the second at index 1, and so on.

Let's begin with the stacked `sales_data` matrix from the previous example. This contains the sales vectors from 2024 and 2025.

```
import numpy as np

# Sales quantities (vectors) for 3 products
q_2024 = np.array([100, 200, 150])
q_2025 = np.array([120, 210, 160])

# Stack the vectors as columns
Q = np.column_stack([q_2024, q_2025])

print("--- Sales Matrix (Q) ---")
print(Q)
```

Again, NumPy uses an intuitive bracket notation to select data.

```
matrix[row_selector, column_selector]
```

To get a single value, you simply provide its coordinates. For example, to find the sales for "Product 1" (row 1) in "2025" (column 1), you would use Q[1, 1], which would return 210.

The real power comes from using the colon (:) as a wildcard, which means "select all." If you want to get the entire sales history for "Product 0" (row 0), you ask for row 0 and all columns. The syntax for this is Q[0, :]. This returns the full vector [100, 120], which you can then analyze or plot.

This works the same way for columns. To isolate the sales for all products in just "2024" (column 0), you would use Q[:, 0]. This would give you the vector [100, 200, 150], which is the original q_2024 vector. This ability to instantly pull out a single year's data or a single product's history is a fundamental part of daily data analysis. Mastering this [row, column] slicing is the key to moving from looking at a whole dataset to surgically extracting the exact insights you need.

Matrix Multiplication

There are many cases where you want to multiply a matrix by another, or perform *matrix multiplication*. When you multiply a price vector by the sales matrix, you compute revenue for multiple years at once. The following example multiplies the prices vector by the Q sales matrix:

```
import numpy as np

# Sales quantities (vectors)
q_2024 = np.array([100, 200, 150])
q_2025 = np.array([120, 210, 160])

Q = np.column_stack([q_2024, q_2025])
prices = np.array([5, 10, 15])

# Revenue for 2024 and 2025
revenues = prices @ Q
print(revenues)
```

Running this snippet produces the following result:

```
[4750 5100]
```

What did the @ operator actually do? It took the prices vector and performed a dot product against every single column of the Q matrix. With one simple operation, you calculated the total revenue for 2024 ($4,750) and 2025 ($5,020). Note the following:

➤ **The first element (4750):** This is the result of np.dot(prices, q_2024), which is $(5 \times 100) + (10 \times 200) + (15 \times 150)$.

➤ **The second element (5020):** This is the result of np.dot(prices, q_2025), which is $(5 \times 120) + (10 \times 210) + (15 \times 160)$.

This ability to perform operations on entire datasets at once is why linear algebra is the backbone of data science and analytics. Instead of writing a for loop to calculate revenue for each year, you can express your problem as a single matrix operation. This ability to perform batch operations on entire datasets at once makes the code faster, cleaner, and more powerful.

Transpose

The transpose flips rows into columns and columns into rows. Sometimes you'll want to change perspective: from "sales by product across years" to "sales by year across products." That can be done by simply using the transpose attribute on a matrix, which is done by adding `.T`, as shown here:

```python
import numpy as np

# Sales quantities (vectors) for 3 products
q_2024 = np.array([100, 200, 150])
q_2025 = np.array([120, 210, 160])

# Stack the vectors as columns
Q = np.column_stack([q_2024, q_2025])

print(Q.T)
```

The result is as follows:

$$\begin{bmatrix} 100 & 200 & 150 \\ 120 & 210 & 160 \end{bmatrix}$$

Transposing a matrix is rotating the data, making it easier to view from different perspectives. But it can also be important for mathematical operations later.

Linear algebra is powerful because it mirrors the way businesses already think, but with the precision and efficiency of mathematics. With Python, these operations are not abstract ideas but real, computable processes that managers, analysts, and students can run instantly.

CREATING AND MANIPULATING VECTORS (AND MATRICES) WITH NUMPY

Let's put these linear algebra skills to the test with a classic finance problem: creating and tracking a stock portfolio. You'll use NumPy to see how these concepts make complex financial analysis surprisingly straightforward.

The goal is to model a portfolio of three stocks over a six-day period. You will:

1. Start with a matrix of historical prices.

2. Calculate a matrix of daily stock returns from those prices.

3. Test two different investment strategies by combining the returns with weight vectors.

The two strategies you'll compare are as follows:

➤ A *constant-weight* "set-it-and-forget-it" approach.

➤ A *time-varying* approach that actively rebalances the portfolio daily.

First, you need data to work with. This example creates a price matrix called `stock_prices` for three fictional stocks ("A", "B", and "C") over six trading days. In this matrix, each row will represent a day, and each column will represent a stock. This arrangement, often described by its "shape" (number of rows, number of columns), is a standard convention in financial analysis. Listing 4-1 sets up all the initial data: the stock tickers, the dates, and the `stock_prices` matrix.

LISTING 4-1: SETTING UP THE DATA

```
import numpy as np

# Setup: tickers, dates, and a small price matrix
tickers = np.array(["A", "B", "C"])
dates = np.array(["2025-01-02","2025-01-03","2025-01-06","2025-01-07","2025-01-
08","2025-01-09"])

# Price matrix: rows are dates, columns are assets
# A     B     C
stock_prices = np.array([
    [100.,   50.,   25.],
    [101.,   49.,   25.5],
    [102.,   48.,   26.0],
    [101.,   48.5, 25.8],
    [103.,   49.,   25.9],
    [104.,   50.,   26.2]
])

# Get the dimensions of the matrix using the .shape attribute
# .shape returns a tuple: (number_of_rows, number_of_columns)
num_time_periods, num_assets = stock_prices.shape

print("Prices matrix shape:", stock_prices.shape)
print(f" (This means {num_time_periods} time periods and {num_assets} assets)")
print("Assets:", tickers)
print("Dates:", dates)
```

The result of the setup is a matrix, P, of stock returns, with asset names and dates:

```
Prices matrix shape: (6, 3)
Assets: ['A' 'B' 'C']
Dates: ['2025-01-02' '2025-01-03' '2025-01-06' '2025-01-07' '2025-01-08'
 '2025-01-09']
```

Running this code block sets up your environment. The first output line, `Prices matrix shape:` `(6, 3)`, confirms the dimensions of the `stock_prices` matrix. The `.shape` function gives you the dimensions of a matrix. This is a key check: as you can see from the new `print` statement, you have six rows (the `num_time_periods`) and three columns (the `num_assets`), just as designed.

The subsequent lines print the asset tickers and the corresponding dates. With the raw price data now loaded into a clean NumPy matrix, you are ready to begin the analysis and calculate the daily returns.

Step 1: Compute Asset Returns from Prices

A stock's price is just a number; its performance is measured by its return. For analysis, you'll calculate the simple daily return, which answers the question, "What was the percentage change from yesterday to today?" The formula is as follows:

```
(today's price / yesterday's price) - 1
```

You could, of course, write a `for` loop to iterate over every day and every stock to calculate this, but NumPy gives you a much more powerful and elegant way to do this for all stocks and all days in a single line. This is done using a feature called *broadcasting*.

Broadcasting is NumPy's function for performing arithmetic operations on arrays of different shapes. It "stretches" the smaller array to match the shape of the larger one without the overhead of actually copying the data. You can use a bit of array slicing to make this work. If you want to divide today's prices by yesterday's prices, you can create two new matrices from the `stock_prices` matrix: one that represents today and one that represents yesterday. `stock_prices[1:]` is a slice of the price matrix `stock_prices` containing all rows from the second day to the end. This will be your today's price matrix. `stock_prices[:-1]` is a slice containing all rows from the first day up to (but not including) the last day. This will be your yesterday's price matrix.

Because these two new matrices have the exact same shape (five rows and three columns), NumPy can perform element-wise division on them instantly. As shown in Listing 4-2, you then subtract 1 to get the final return. The resulting matrix, named `Returns`, will naturally have one fewer row than your original `stock_prices` matrix, because returns only occur between days.

LISTING 4-2: PERFORMING ELEMENT-WISE DIVISION

```
import numpy as np
stock_prices = np.array([
    [100.,  50.,   25.],
    [101.,  49.,   25.5],
    [102.,  48.,   26.0],
    [101.,  48.5,  25.8],
    [103.,  49.,   25.9],
    [104.,  50.,   26.2]
])

returns = stock_prices[1:] / stock_prices[:-1] - 1
```

Now, each row of R is a vector of asset returns for that period. This is exactly the "vectors of numbers" view you want: a period's returns are just a return vector. Stack those vectors, and you have a return matrix.

To see how the three stocks performed, you can import the `pyplot` library from Matplotlib and create a simple visualization of your new returns matrix, `returns`. This is shown in Listing 4-3.

LISTING 4-3: SEEING HOW THE THREE STOCKS PERFORMED

```
import matplotlib.pyplot as plt
import numpy as np
stock_prices = np.array([
    [100.,   50.,   25.],
    [101.,   49.,   25.5],
    [102.,   48.,   26.0],
    [101.,   48.5, 25.8],
    [103.,   49.,   25.9],
    [104.,   50.,   26.2]
])

returns = stock_prices[1:] / stock_prices[:-1] - 1

plt.plot(R)
plt.show()
```

The result of this plot is shown in Figure 4-1.

Step 2: Portfolio with Constant Weights

The first strategy for your financial problem is simple: you allocate your money at the start and never change it. This is the "set-it-and-forget-it" approach.

To model this, this example creates a weight vector called `weight` that assigns 50% of your portfolio to Stock A, 30% to B, and 20% to C. To find the portfolio's total return on any given day, you just need to calculate the dot product of that day's return vector (a row from `returns`) and your w vector.

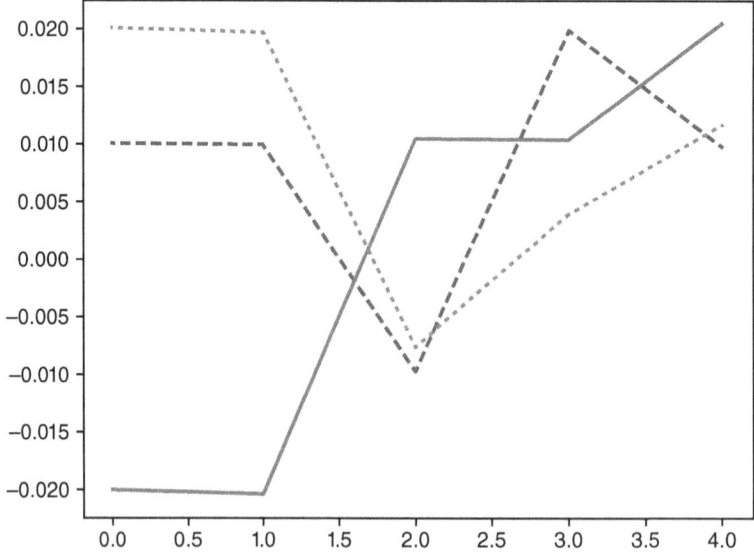

FIGURE 4-1: A simple plot showing stock returns over time.

Thanks to NumPy, you can calculate the entire time series of portfolio returns in one shot using matrix-vector multiplication, `returns @ weights`. Listing 4-4 shows the full listing of this solution.

LISTING 4-4: CALCULATING WITH MATRIX-VECTOR MULTIPLICATION

```python
import numpy as np
stock_prices = np.array([
    [100.,  50.,   25.],
    [101.,  49.,   25.5],
    [102.,  48.,   26.0],
    [101.,  48.5, 25.8],
    [103.,  49.,   25.9],
    [104.,  50.,   26.2]
])


returns = stock_prices[1:] / stock_prices[:-1] - 1

# Constant weights for our three stocks (must sum to 1)
weights = np.array([0.5, 0.3, 0.2])

# Calculate the portfolio's return for every single day
# This is matrix-vector multiplication: (5, 3) @ (3,) -> (5,)
r_p_const = returns @ weights

# Let's see how $1 invested at the start would have grown
# We need to calculate the cumulative product of the daily growth
portfolio_index_const = np.cumprod(1 + r_p_const)

print("Ending portfolio value (constant weights):", portfolio_index_const[-1])
```

The `returns @ weights` operation efficiently computes the portfolio's performance over time. This is the same math as (weight_A × return_A) + (weight_B × return_B) + ⋯, repeated for every day. The resulting portfolio value with constant weights is `1.029`.

To track the growth of a $1 investment, you use the `np.cumprod` function. This is a powerful tool for calculating a cumulative product. It takes your series of daily returns (e.g., [1.003, 1.002, 0.999]) and calculates the compounding growth: [1.003, (1.003 × 1.002), (1.003 × 1.002 × 0.999)]. This gives the index value for each day. You can also visualize the portfolio value over time using the code in Listing 4-5.

LISTING 4-5: VISUALIZING THE PORTFOLIO VALUE OVER TIME

```python
import numpy as np
import matplotlib.pyplot as plt

stock_prices = np.array([
    [100.,  50.,   25.],
    [101.,  49.,   25.5],
    [102.,  48.,   26.0],
    [101.,  48.5, 25.8],
```

```
        [103.,  49.,   25.9],
        [104.,  50.,   26.2]
    ])

returns = stock_prices[1:] / stock_prices[:-1] - 1

# Constant weights for our three stocks (must sum to 1)
weights = np.array([0.5, 0.3, 0.2])

# Calculate the portfolio's return for every single day
# This is matrix-vector multiplication: (5, 3) @ (3,) -> (5,)
r_p_const = returns @ weights

# Let's see how $1 invested at the start would have grown
# We need to calculate the cumulative product of the daily growth
portfolio_index_const = np.cumprod(1 + r_p_const)
plt.plot(portfolio_index_const)
plt.show()
```

The results of this plot are shown in Figure 4-2.

Step 3: Portfolio with Time-varying Weights

Most serious strategies involve periodic rebalancing. The "set-it-and-forget-it" model was simple, but this step models a simple *active* strategy. Instead of a single weight *vector*, you will create a weight *matrix* (called WEIGHTS), where each row represents a different weight vector for each day.

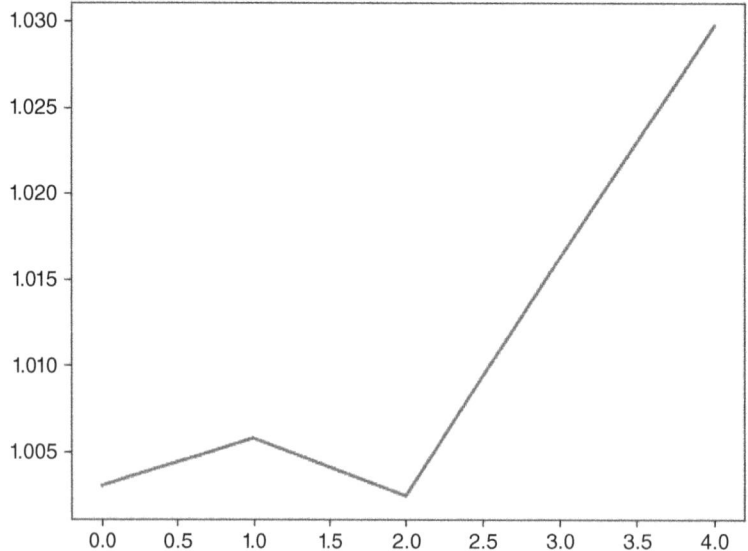

FIGURE 4-2: A simple line plot showing portfolio value over time.

The simple rule: each day, you'll slightly increase your investment in whichever stock performed best the day before.

To set this up, you first create a 5 × 3 W matrix by "tiling" the constant `weights` vector. To do this, you use `np.tile`, which is a function that repeats an array. This function is a "tiling" utility: it takes an array and repeats it. You're telling NumPy to take your `weights` vector and tile it five times vertically (`returns.shape[0]`) and one time horizontally, producing a 5 × 3 matrix of the starting weights.

Then, you'll loop from the second day onward. Inside the loop, you use `np.argmax` to find the index (0, 1, or 2) of yesterday's (t – 1) winning stock. You then nudge the weight for that winner up by 0.05 and re-normalize the row so all weights sum back to 1. This is all shown in Listing 4-6.

LISTING 4-6: USING A WEIGHT MATRIX

```python
import numpy as np
import matplotlib.pyplot as plt

stock_prices = np.array([
    [100.,   50.,   25.],
    [101.,   49.,   25.5],
    [102.,   48.,   26.0],
    [101.,   48.5, 25.8],
    [103.,   49.,   25.9],
    [104.,   50.,   26.2]
])

returns = stock_prices[1:] / stock_prices[:-1] - 1

# Constant weights for our three stocks (must sum to 1)
weights = np.array([0.5, 0.3, 0.2])
# Start with our constant weights as a baseline
WEIGHTS = np.tile(weights, (returns.shape[0], 1))

# A simple rebalancing rule:
for t in range(1, returns.shape[0]):
    # Find the index of yesterday's winning stock
    winner_idx = np.argmax(returns[t-1])

    # Nudge the weight for today's portfolio toward that winner
    WEIGHTS[t, winner_idx] += 0.05

    # Re-normalize the row to ensure weights sum back to 1
    WEIGHTS[t] = WEIGHTS[t] / WEIGHTS[t].sum()

# Calculate time-varying portfolio returns
# This is a row-by-row dot product: (Day 1 returns * Day 1 weights) + ...
r_p_timevary = np.sum(returns * WEIGHTS , axis=1)

portfolio_index_timevary = np.cumprod(1 + r_p_timevary)

print("Ending portfolio value (time-varying weights):", portfolio_index_time-
vary[-1])
```

The for loop—`for t in range(1, returns.shape[0]):`—is where the new logic lives. Notice that it starts at 1 (the second day), not 0, because this rule depends on yesterday's (`t-1`) performance.

Inside the loop, the first line is `winner_idx = np.argmax(returns[t-1])`. The `np.argmax` function is a new and incredibly useful tool. It doesn't return the highest value from the array; instead, it returns the index (0, 1, or 2) of that highest value. This is exactly what you need. You're asking it, "For yesterday's returns, which stock won?" This returns 0 (for Stock A), 1 (for Stock B), or 2 (for Stock C).

The next line, `WEIGHTS[t, winner_idx] += 0.05`, uses that index. It's a precise update: "Go to the `WEIGHTS` matrix, find the row for today (t), and in that row, find the column for `winner_idx` and add 0.05 to its weight."

Of course, this means your weights for that day no longer add up to 1. That's why the final line in the loop, `WEIGHTS[t] = WEIGHTS[t] / WEIGHTS[t].sum()`, is so critical. It's a "re-normalization" step. It calculates the new sum of the row and then divides the entire row by that sum. This scales all weights back down, ensuring they once again add up to 1.

After the loop is finished, you have your `WEIGHTS` matrix, where each row is a slightly different portfolio allocation. The `r_p_timevary = np.sum(returns * WEIGHTS, axis=1)` line is the NumPy way of performing a row-by-row dot product. The `returns * WEIGHTS` operation performs an element-wise multiplication, and `np.sum(..., axis=1)` then sums those products horizontally, giving the total portfolio return for each specific day. Finally, you can use `np.cumprod` just as before to compute the compounded growth of `1.03`. You can also visualize the results with a plot, as shown in Figure 4-3.

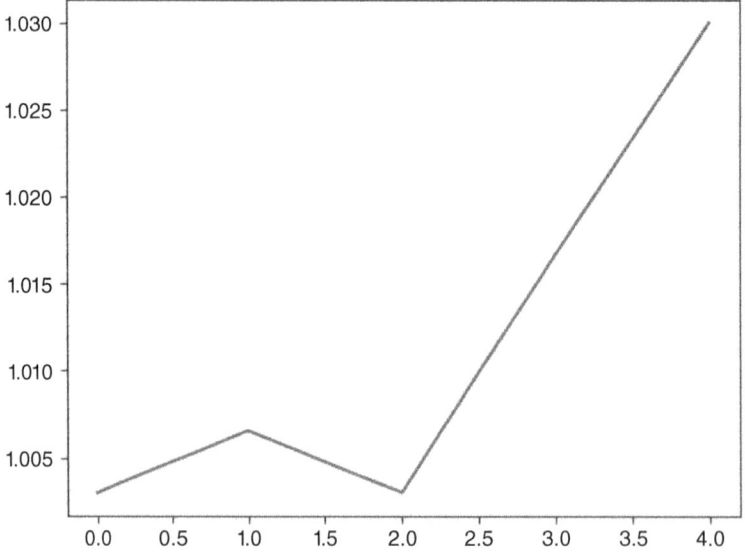

FIGURE 4-3: A line plot showing the results of using time-varying weights.

Comparing Strategies (Same Math, Different Inputs)

Because you used the same linear algebra framework for both the constant weights strategy as well as for time-varying weights strategy, the final results are two clean vectors that you can easily compare. Consider for example how an initial $1 investment would have performed with each strategy side-by-side.

Listing 4-7 uses extra print statements to display the results in a visual manner. You'll use the enumerate function in the loop. enumerate is a helpful Python function that returns two things at once: the index number (i) and the item from the list (date). This enables you to use i to access your portfolio data while using date to print the corresponding date. Before running Listing 4-7, run the code from Listing 4-6 (it defines the data you will use for the visualization).

LISTING 4-7: EXTRA PRINT STATEMENT TO DISPLAY RESULTS VISUALLY

```python
### Include Listing 4.6 code here ###
import numpy as np
import matplotlib.pyplot as plt

print("--- Performance Comparison (Index Value) ---")
print("Date        | Constant | Time-Varying")
print("----------------------------------------")

for i, date in enumerate(dates[1:]):
    const_val = portfolio_index_const[i]
    timevary_val = portfolio_index_timevary[i]
    print(f"{date} | {const_val:.4f}   | {timevary_val:.4f}")

plt.plot(portfolio_index_const, label="Constant")
plt.plot(portfolio_index_timevary, label="Time-Varying")
plt.legend()
plt.show()
```

The results of this listing are the following text along with the plot presented in Figure 4-4.

```
--- Performance Comparison (Index Value) ---
Date         | Constant | Time-Varying
----------------------------------------
2025-01-03 | 1.0030   | 1.0030
2025-01-06 | 1.0058   | 1.0066
2025-01-07 | 1.0024   | 1.0030
2025-01-08 | 1.0162   | 1.0167
2025-01-09 | 1.0297   | 1.0300
```

Notice how the same core operations—dot products and matrix multiplications—allow you to model two distinct financial strategies. This is the beauty of linear algebra: it provides a powerful and efficient language for describing complex systems.

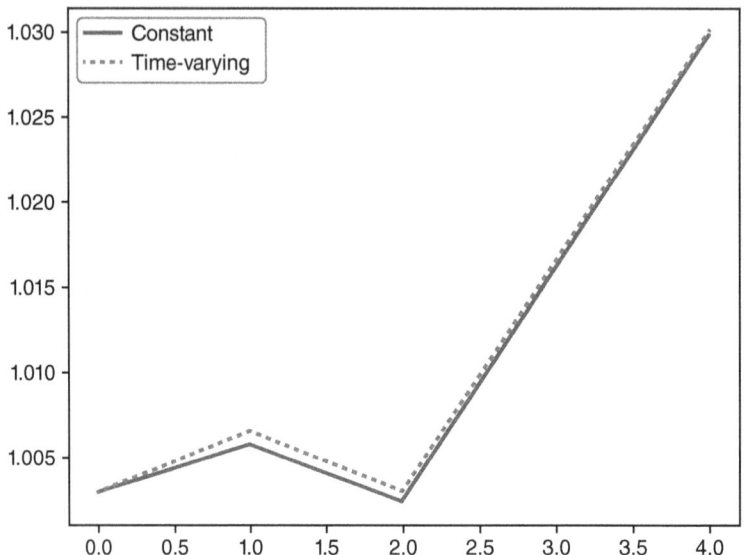

FIGURE 4-4: Performance of both the constant and time-varying weights over time.

EIGENVALUES AND EIGENVECTORS: BUSINESS APPLICATIONS

Beyond day-to-day operations, every business system has dominant trends, stable states, and long-term tendencies. Eigenvalues and eigenvectors are the advanced tools that help you discover this hidden engine.

While the names sound complex, the idea is surprisingly intuitive. They help you answer critical questions like these:

➤   If current trends continue, where will our customer base be in a year?

➤   What is the single biggest risk factor driving our investment portfolio?

➤   Who are the most influential players in our supply chain network?

What Eigenvalues and Eigenvectors Represent

A matrix acts like a "wind machine," applying a force that transforms vectors. When this "wind" hits most vectors, it blows them in a completely new direction. But some special vectors, the *eigenvectors*, are perfectly aligned with the wind. When the wind hits them, they don't change direction. They only get stretched, shrunk, or flipped. The factor by which they stretch or shrink is the *eigenvalue*.

In business, you can think of eigenvectors as the natural directions or stable trends of a system, and eigenvalues as the strength or magnitude of those trends.

Formally, for a matrix A and its eigenvector v, the relationship is as follows:

$$A\theta = \lambda\theta$$

Where λ (lambda) is the eigenvalue.

Why Eigenvalues Matter for Long-term Stability

Eigenvalues are crucial for understanding whether a system will grow, shrink, or stabilize over time. Specifically, the value of the system's dominant eigenvalue dictates its long-term behavior:

➤ If the dominant eigenvalue is greater than 1, the system grows exponentially. This could be viral marketing (good) or compounding portfolio risk (bad).

➤ If the dominant eigenvalue is less than 1, the system decays and fades away over time.

➤ If the dominant eigenvalue is exactly 1, the system converges to a stable, steady state, like a fixed market share or a predictable distribution of loyal customers.

To see this in action, you'll see two powerful examples. First, you'll see how an eigenvector can predict long-term customer loyalty and see if your business model has a "leaky bucket." Second, you'll use eigenvectors to analyze a stock portfolio and uncover the single biggest risk factor driving its volatility.

Predicting Long-term Customer Loyalty

Imagine you're tracking your customers, who can be in one of three states: Active, Dormant, or Churned. You create a transition matrix that shows the probability of a customer moving from one state to another each month.

The business question: If you do nothing, where will all your customers eventually end up? The answer lies in the eigenvector associated with the eigenvalue of 1, which will give you the system's steady state, which you can calculate in the code presented in Listing 4-8.

LISTING 4-8: USING AN EIGENVECTOR

```
import numpy as np

# A customer transition matrix
# Rows: Current State, Columns: Next State
# States: [Active, Dormant, Churned]
P = np.array([
    [0.7, 0.2, 0.1],
    [0.3, 0.5, 0.2],
    [0.0, 0.0, 1.0]
])
```

```
# To find the steady-state for this type of problem, we use the transpose P.T
eigenvalues, eigenvectors = np.linalg.eig(P.T)

# Find the eigenvector corresponding to the eigenvalue of 1
steady_state_vector = eigenvectors[:, np.isclose(eigenvalues, 1)]
steady_state_vector = steady_state_vector[:, 0].real
steady_state_vector = steady_state_vector / steady_state_vector.sum()

print("Eigenvalues:", eigenvalues)
print("Steady-State Distribution (A, D, C):", steady_state_vector)
```

Running this code will produce the following output:

```
Eigenvalues: [1.         0.33542487 0.86457513]
Steady-State Distribution (A, D, C): [0. 0. 1.]
```

The first thing to notice is that it's analyzing `np.linalg.eig(P.T)`, which is the eigenvector of the transpose of the matrix. For finding the steady state of a transition matrix, the math requires you to find the "left eigenvector," which, for the purposes here, you can get by finding the regular eigenvector of the transposed matrix. The `np.linalg.eig` function gives two arrays: eigenvalues and eigenvectors. The raw eigenvalues are [1., 0.5, 0.2], which confirms the system has a stable end-state because the dominant eigenvalue is exactly 1.

The next step is to find the eigenvector that corresponds to that eigenvalue of 1. The `np.isclose(eigenvalues, 1)` line creates a Boolean mask ([True, False, False]) that finds the index of the eigenvalue equal to 1. You use this mask with the slicing syntax `eigenvectors[:, ...]` to pull out the correct column (the eigenvector) from the eigenvectors matrix. This raw vector isn't clean; it might be [0., 0., 0.89442719]. To make it a proper probability, you use `.real` to get only its real part (ignoring any tiny complex numbers from computation) and then normalize it by dividing the vector by its own sum. This scales all the numbers so they add up to 1.

The final, normalized `steady_state_vector` tells a stark story: [0. 0. 1.]. This means that, eventually, 0% of customers will be Active, 0% will be Dormant, and 100% will be Churned. The results from the eigenvector analysis show that you customers are a leaky bucket, which means that every customer is destined to leave.

Uncovering the Biggest Risk in a Stock Portfolio

In finance, an asset's risk is not just about its own volatility, but how it moves with other assets. A covariance matrix captures this interconnected risk, showing, for example, if Stock A tends to go up when Stock B goes down. NumPy provides a simple function, called `np.cov`, to calculate this for you. It takes a matrix of data (where each row is a variable and each column is an observation) and returns a new matrix showing the covariance between all the variables.

By finding the eigenvalues and eigenvectors of this matrix, you can perform a simplified version of Principal Component Analysis (PCA), as shown in Listing 4-9, to answer the question: What is the single biggest source of risk driving my entire portfolio? Let's start with a simple matrix of daily returns for three stocks. This matrix will have days as rows and stocks as columns.

LISTING 4-9: A SIMPLIFIED VERSION OF PRINCIPAL COMPONENT ANALYSIS

```python
import numpy as np

# A simple matrix of daily returns for three stocks
# Rows = days, Columns = stocks
returns = np.array([
    [0.01,  0.02,   0.015],
    [0.005, 0.018,  0.010],
    [-0.002,-0.015, -0.008],
    [0.012, 0.020,  0.017],
])

# 1. Compute the covariance matrix
cov_matrix = np.cov(returns.T)

# 2. Get eigenvalues and eigenvectors (eigh is best for symmetric matrices)
eigvals, eigvecs = np.linalg.eigh(cov_matrix)

# 3. Find the principal component (the one with the largest eigenvalue)
idx = np.argmax(eigvals)
principal_eigenvalue = eigvals[idx]
principal_eigenvector = eigvecs[:, idx]

print(f"Largest Eigenvalue (Magnitude of Risk): {principal_eigenvalue:.6f}")
print(f"Principal Eigenvector (The 'Market Factor'): {principal_eigenvector}")
```

Running this code will produce the following output:

```
Largest Eigenvalue (Magnitude of Risk): 0.000455
Principal Eigenvector (The 'Market Factor'): [-0.27500973 -0.80216434
-0.5300019 ]
```

First, it used `np.cov(R.T)`. You need to provide the transpose of your `returns` matrix because `np.cov` expects each row to be a variable and each column to be an observation.

Next is `np.linalg.eigh(cov_matrix)`. This function is similar to `np.linalg.eig`, but it's specifically designed for symmetric matrices (where the matrix is a mirror image of itself across the diagonal). A covariance matrix is always symmetric, and using `eigh` is often faster and more numerically stable. The output shows three eigenvalues. The `np.argmax(eigvals)` function found the index of the largest one, `0.000455`. This largest eigenvalue represents the variance (or "strength") of the most dominant risk factor.

The corresponding principal eigenvector is `[-0.27500973 -0.80216434 -0.5300019]`. This is a recipe showing how each stock is exposed to this main risk driver. Because all three values are positive, it suggests this dominant factor is a broad market risk that tends to move all three stocks in the same direction. Analysts use this technique to distill the risk of thousands of assets down to a handful of key driving factors.

SUMMARY

Throughout this chapter, you have journeyed from the fundamental building blocks of data, vectors and matrices, to the profound insights offered by eigenvectors. More than just a collection of mathematical rules, you've discovered that linear algebra provides a powerful new lens for viewing business problems. With Python and NumPy, this chapter bridged the gap between theory and practice, turning abstract concepts into concrete solutions, such as calculating revenues, modeling portfolio performance, and even predicting the long-term stability of a system. As you move forward, see these tools not as abstract requirements, but as a core part of your analytical toolkit. The ability to structure a problem in terms of vectors and matrices is the first step toward solving some of the most complex and rewarding challenges in business and data science.

CONTINUE YOUR LEARNING

This chapter used several powerful NumPy functions to bring the concepts of linear algebra to life. Table 4-1 serves as a quick reference, summarizing the key vector/matrix manipulation tools.

TABLE 4-1: NumPy's Manipulation Tools

FUNCTION NAME	NUMPY FUNCTION	DESCRIPTION
Dot product	np.dot(a, b)	Calculates the dot product of two vectors or the product of two matrices.
Matrix multiplication (shortcut)	A @ b	A shorthand operator for matrix multiplication (np.matmul).
Transpose	np.transpose(a) or a.T	Flips a matrix over its diagonal, turning its rows into columns and its columns into rows.
Array (creation)	np.array([...])	The fundamental function to create a NumPy array (vector or matrix) from a list.
Column stack	np.column_stack([...])	Takes a list of 1D vectors and stacks them together as columns to form a 2D matrix.
Covariance	np.cov(m)	Calculates the covariance matrix, showing how variables move together.

TABLE 4-2: NumPy's Linear Algebra Functions

FUNCTION NAME	NUMPY FUNCTION	DESCRIPTION
Vector norm (magnitude)	`np.linalg.norm(a)`	Calculates the L2 norm (Euclidean length or magnitude) of a vector.
Eigenvalues/eigenvectors	`np.linalg.eig(a)`	Computes the eigenvalues and eigenvectors of a square matrix.
Eigenvalues/eigenvectors (symmetric)	`np.linalg.eigh(a)`	A specialized, more stable version of `eig` for symmetric matrices (like a covariance matrix).
Solve linear system	`np.linalg.solve(A, b)`	Solves a system of linear equations in the form `Ax = b`.
Matrix inverse	`np.linalg.inv(a)`	Calculates the inverse of a square matrix.
Find max index (arg max)	`np.argmax(a)`	Returns the *index* (not the value) of the maximum element in an array.
Tile array (repeat)	`np.tile(A, reps)`	"Tiles" or repeats an array `A` a specified number of times.

In addition, while NumPy contains hundreds of functions, a handful are essential for day-to-day data analysis and linear algebra. Table 4-2 lists the core functions used in this chapter, which are split between NumPy's main library and its specialized `linalg` (linear algebra) submodule.

5

Calculus for Business Problem Solving

From just living your daily life, you are probably already an expert in the study of change. In business, we track whether sales are rising or falling, if customer growth is accelerating, and when costs are starting to creep up. This constant motion is the lifeblood of any enterprise. But what if you had a precise mathematical language to describe, measure, and even predict these dynamics? That language is calculus.

In the business world, calculus is a practical toolkit for understanding the momentum of your operations. It's the difference between knowing your company's yearly revenue and knowing the exact rate at which that revenue was growing on the day you launched a new marketing campaign. It allows you to pinpoint the exact moment when your marketing efforts hit the point of diminishing returns, and it gives you a framework for modeling the entire lifecycle of a new product, from launch to market saturation.

This chapter demystifies calculus by putting it to work on real-world business problems using Python. It starts with common business data, like daily sales figures, and uses numerical calculus to find rates of change, acceleration, and total accumulation. The Python ecosystem is rich with tools to do calculus and so you'll take a brief tour of the Python calculus ecosystem, from the numerical power of NumPy and SciPy to the symbolic capabilities of SymPy. From there, the chapter moves into forecasting, using differential equations to model growth curves. You'll even tackle multi-variable problems by visualizing a "profit landscape" and exploring how to measure the impact of individual decisions. Finally, the chapter brings all these concepts together in a case study that uses marginal analysis to understand the dynamics of profitability.

By the end of this chapter, you'll be equipped to use these powerful mathematical ideas in your Python code. You will move from simply observing data to actively analyzing the dynamics behind it, laying the foundation for making smarter, more data-driven business decisions.

NUMERICAL DIFFERENTIATION AND INTEGRATION IN BUSINESS ANALYTICS

In a perfect textbook world, business trends would follow smooth, predictable curves that we could define with elegant functions. In reality, data is usually sparse and noisy. For example, consider a series of discrete sales measurements taken day by day or month by month. This is where the power of numerical calculus comes in. You can apply the core ideas of differentiation and integration directly to your data to uncover powerful insights about rates of change, pinpoint the exact moment of diminishing returns, and calculate total accumulation, all without needing a perfect formula. The following sections explore each of these concepts in detail: using the derivative to find the rate of change, the second derivative to find the point of diminishing returns, and the integral to accumulate totals.

The Derivative: Finding the Rate of Change

The derivative is simply a measurement of how a value is changing at a specific point in time. Think of it as the speedometer for your business. Your car's main odometer tells you the total distance you've traveled (your cumulative sales), but it's the speedometer that tells you how fast you're going right now (your sales growth rate). This instantaneous speed is crucial for making timely decisions.

Consider this business question: You just launched a new mobile game, and you're tracking the total number of downloads. Is the excitement for your game speeding up or slowing down? The total download count will always go up, but the derivative (the rate of new downloads per day) tells you about the project's momentum.

In Python, you don't need complex formulas to answer this question. When working with a sequence of data points (like daily sales figures), you can use NumPy's diff() function to find the rate of change between consecutive data points. This gives you a practical, powerful way to see how the data is changing over time.

Imagine your game's download numbers over its first 10 days. As shown in Listing 5-1, you can store this data as a NumPy array. Then you can use NumPy's .diff() function to compute the difference.

LISTING 5-1: ANALYZING MOBILE GAME DOWNLOADS

```python
import numpy as np
import matplotlib.pyplot as plt

# Daily cumulative downloads for the first 10 days
cumulative_downloads = np.array([0, 150, 400, 800, 1300, 1850, 2400, 2900, 3300,
3600])

# Calculate the daily new downloads using np.diff()
daily_new_downloads = np.diff(cumulative_downloads)

# Print the results
print(f"Cumulative Downloads: {cumulative_downloads}")
print(f"Daily New Downloads:  {daily_new_downloads}")
```

```
# --- Visualization ---
fig, ax1 = plt.subplots(figsize=(10, 6))

# Plot cumulative downloads on the primary y-axis (ax1)
color = 'tab:blue'
ax1.set_xlabel('Day')
ax1.set_ylabel('Total Downloads', color=color)
ax1.plot(cumulative_downloads, color=color, marker='o', line-
style='-', label='Cumulative Downloads')
ax1.tick_params(axis='y', labelcolor=color)
ax1.grid(True)

# Create a secondary y-axis (ax2) for the daily rate
ax2 = ax1.twinx()
color = 'tab:green'
ax2.set_ylabel('New Downloads per Day', color=color)
# We plot from day 1, as the rate is between days
ax2.plot(range(1, len(daily_new_downloads) + 1), daily_new_down-
loads, color=color, marker='s', linestyle='--', label='Daily New Downloads')
ax2.tick_params(axis='y', labelcolor=color)

fig.suptitle('Mobile Game Download Momentum', fontsize=16)
fig.tight_layout()
plt.show()
```

Listing 5-1 starts by creating an array, `cumulative_downloads`, that holds the daily cumulative download values. In a real-world program, this data could be read from a file rather than hardcoded. This data array is then passed to the NumPy `.diff()` function. This function computes the difference for the full array. It effectively subtracts Day 1 from Day 2, Day 2 from Day 3, and so on, for every column simultaneously. The computed value provides a new array holding the daily new download values. The results of these lines are then printed to the screen:

```
Cumulative Downloads: [0 150 400 800 1300 1850 2400 2900 3300 3600]
Daily New Downloads: [150 250 400 500 550 550 500 400 300]
```

After printing the values, the program creates Figure 5-1 to plot both datasets on a single graph. This is done using Matplotlib's `twinx()` feature, which allows you to have two different y-axes sharing the same x-axis (days).

First, you set up the primary y-axis (`ax1`) to plot the `cumulative_downloads`. Use a solid line (`linestyle='-'`) with circular markers (`marker='o'`) to represent the steady, upward trend of total downloads.

Next, you create a secondary y-axis (`ax2 = ax1.twinx()`) for the daily rate. This is crucial because the scale of daily new downloads (hundreds) is much smaller than the scale of cumulative downloads (thousands). If you plotted them on the same axis, the daily rate line would look flat and insignificant at the bottom of the chart. By giving it its own axis on the right side, you can see its true shape clearly. You then plot this `daily_new_downloads` data using a dashed line (`linestyle='--'`) with square markers (`marker='s'`) to distinguish it from the cumulative total. Note that since this is plotting the difference in value between two days, it will start on Day 1 instead of Day 0.

The visualization tells a clear story that the raw cumulative numbers hide. As you can see, even though the solid line (Total Downloads) is always rising, the green line (Daily New Downloads)

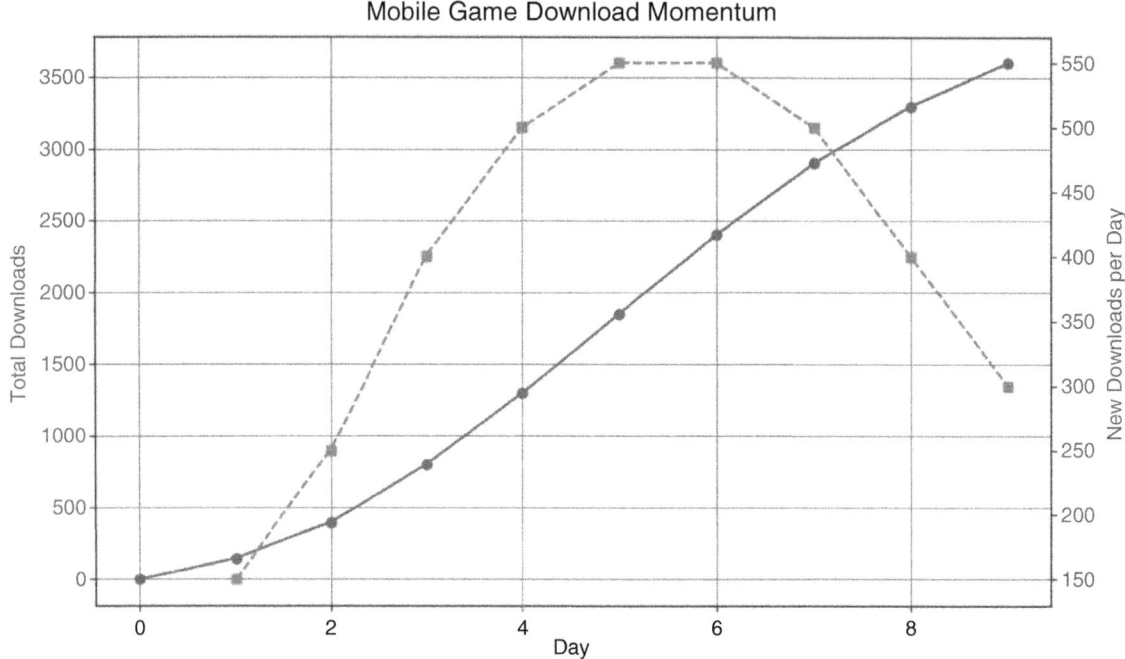

FIGURE 5-1: Total number of mobile game downloads over time.

reveals the true momentum. The game's growth accelerated rapidly for the first five days, peaked when daily downloads were growing at a rate of 550 downloads per day, and has been slowing down since Day 7. This is a critical insight. It tells you that your initial launch buzz is fading and now might be the perfect time for a new marketing push to reaccelerate growth.

The Second Derivative: Pinpointing the Point of Diminishing Returns

The second derivative measures the rate of change of the rate of change. It's not the speed, but the acceleration of your metric. Its most powerful use in business is to find the inflection point—the precise moment when a trend's momentum shifts. For a growing metric, this is the point of diminishing returns: the moment the growth begins to slow down.

To calculate this, you simply take the derivative of the first derivative. In NumPy, this means applying `np.diff()` a second time. Listing 5-2 uses the `daily_new_downloads` array calculated in the previous section and passes it through `np.diff()` again. This gives you a new array, called `download_acceleration`, which represents the day-to-day change in the growth rate. Combine the code from Listing 5-1 and Listing 5-2.

LISTING 5-2: FINDING THE INFLECTION POINT FOR GAME DOWNLOADS

```
# The first derivative is the daily new downloads
daily_new_downloads = np.array([150, 250, 400, 500, 550, 550, 500, 400, 300])
```

```
# The second derivative is the "acceleration" of downloads
download_acceleration = np.diff(daily_new_downloads)

print(f"Download Acceleration: {download_acceleration}")

# --- Visualization ---
plt.figure(figsize=(10, 6))
# Plot the first derivative (daily rate)
plt.plot(range(1, len(daily_new_downloads) + 1), daily_new_downloads, 'g-s',
label='Daily New Downloads (1st Derivative)')
# Plot the second derivative (acceleration)
plt.plot(range(2, len(download_acceleration) + 2), download_acceleration, 'b-^',
label='Download Acceleration (2nd Derivative)')
plt.axhline(y=0, color='grey', linestyle='--')

plt.title('Finding the Point of Diminishing Returns')
plt.xlabel('Day')
plt.ylabel('Change in Downloads')
plt.legend()
plt.grid(True)
plt.show()
```

The result:

```
Download Acceleration: [100 150 100 50 0 -50 -100 -100]
```

The visualization makes the story clear, as you can see in Figure 5-2. The daily rate (squares) and the acceleration (triangles) are plotted on the same graph. Notice that there is also a horizontal dashed line at zero (`plt.axhline(y=0)`). This simple addition is crucial because it acts as a visual boundary.

The lower line (acceleration) shows the momentum. It's positive for the first few days, meaning growth is speeding up. It's zero on Day 6, and then it turns negative. Day 7, where the acceleration first becomes negative (–50), is the inflection point. This is the point of diminishing returns. Even though you still gained 500 new users on Day 7, the growth engine had started to slow down. For a business, this is the signal that a strategy (like an initial marketing blast) has reached its peak effectiveness and it's time to consider a new approach to reaccelerate growth.

The Integral: Accumulating the Totals

Integration is the reverse of differentiation. If the derivative acts as a speedometer by converting the position function into velocity, the integral is the odometer that converts the velocity function into total distance traveled. It allows you to "sum up" a series of small, incremental changes to understand their cumulative impact.

This is incredibly useful in business. Imagine you know the number of new subscribers your streaming service gets each day. How do you find your total subscriber count at the end of the month, by using integration?

For numerical data, this process is wonderfully simple. You can use NumPy's `cumsum()` function, which calculates the cumulative sum of an array. It takes a stream of rate data and returns the total accumulation over time.

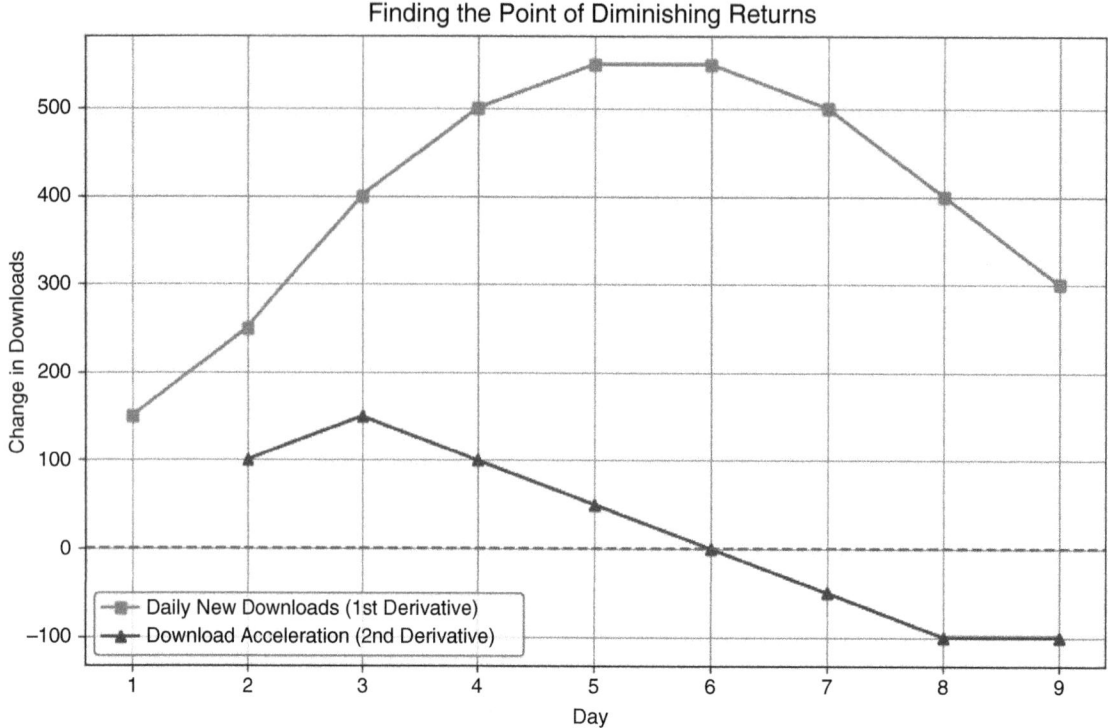

FIGURE 5-2: Visualization of the second derivative.

Say you have the daily numbers for new subscribers over a week. As shown in Listing 5-3, you can use cumsum() to see how the total user base grows.

LISTING 5-3: TRACKING STREAMING SERVICE SUBSCRIBERS

```python
import numpy as np
import matplotlib.pyplot as plt

# Number of new subscribers each day for a week
daily_new_subs = np.array([120, 135, 150, 160, 145, 110, 90])
days = ['Mon', 'Tue', 'Wed', 'Thu', 'Fri', 'Sat', 'Sun']

# Calculate the cumulative sum of subscribers
total_subs = np.cumsum(daily_new_subs)

# Print the results
print(f"Daily New Subscribers: {daily_new_subs}")
print(f"Total Subscribers:     {total_subs}")

# --- Visualization ---
plt.figure(figsize=(10, 6))
plt.bar(days, daily_new_subs, color='skyblue', label='New Subscribers Each Day')
```

```
plt.plot(days, total_subs, color='navy', marker='o', linestyle='-',
label='Total Subscriber Count')

plt.title('Streaming Service Growth Over One Week')
plt.ylabel('Number of Subscribers')
plt.xlabel('Day of the Week')
plt.legend()
plt.grid(axis='y', linestyle='--', alpha=0.7)
plt.show()
```

This code starts with the `daily_new_subs` array. The calculation happens in the single line `total_subs = np.cumsum(daily_new_subs)`. This function iterates through the array, keeping a running total. The first element is just 120. The second element is 120 + 135 = 255. The third is 120 + 135 + 150 = 405, and so on.

The results confirm this:

```
Daily New Subscribers: [120 135 150 160 145 110 90]
Total Subscribers: [120 255 405 565 710 820 910]
```

The visualization code then combines these two views. It uses `plt.bar` to show the daily rate as a light blue bar chart and `plt.plot` to overlay the cumulative total as a dark navy line chart. The plot is shown in Figure 5-3. It clearly shows how the daily rate (the bar chart) directly builds the

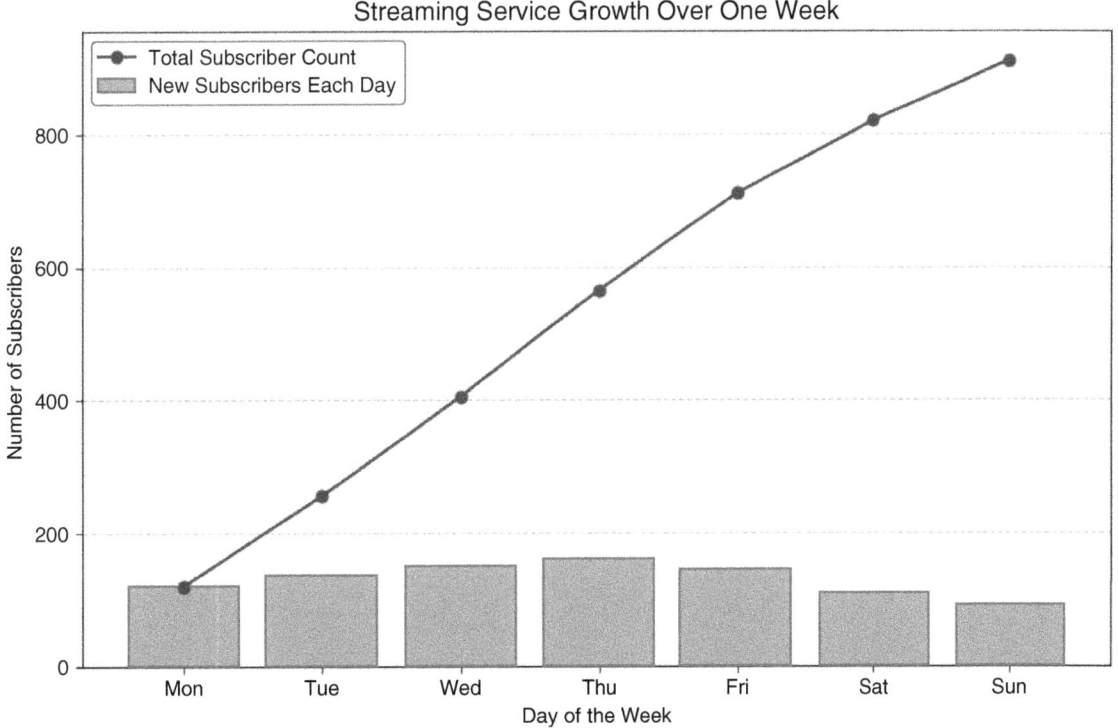

FIGURE 5-3: Demonstration of integration using subscriber growth.

cumulative total (the line chart). You can see that the subscriber acquisition peaked mid-week on Thursday and slowed over the weekend. By the end of the week, those daily gains accumulated to a total of 910 new subscribers. This simple act of integration gives you a complete picture of the daily performance and its overall impact.

THE CALCULUS ECOSYSTEM IN PYTHON

So far, this chapter has focused on the numerical approach to calculus using NumPy, which is perfect when your starting point is a collection of data points. However, the Python ecosystem offers a rich set of tools for different kinds of calculus problems. There are several main approaches, which include numerical calculus, symbolic calculus, and advanced numerical methods.

The following subsections take a brief look at each of these approaches. The discussion starts with numerical calculus using NumPy for analyzing array data. Then, you'll explore symbolic calculus with SymPy for working directly with mathematical formulas. Finally, you'll learn a bit about advanced numerical methods using SciPy for high-precision tasks like integration.

Numerical Calculus with NumPy

Numerical calculus using NumPy is the method you've been using. It's the best choice when you have data in an array and want to find the differences between points (np.diff) or the cumulative sum (np.cumsum). It's fast, efficient, and direct for data analysis.

As an example, imagine a matrix where each row represents a day and each column represents a different product. The values are the cumulative sales for that product. As shown in Listing 5-4, you can use np.diff to find the daily new sales for all products at once.

LISTING 5-4: DIFFERENTIATING A SALES MATRIX

```
import numpy as np

# Cumulative sales matrix: 4 days, 3 products
#        Product A, Product B, Product C
sales_matrix = np.array([
    [100, 50, 200],   # Day 1
    [110, 55, 215],   # Day 2
    [125, 65, 235],   # Day 3
    [145, 80, 250]    # Day 4
])

# Differentiate along the rows (axis=0) to get daily changes
daily_sales_growth = np.diff(sales_matrix, axis=0)

print("--- Daily Sales Growth ---")
print(daily_sales_growth)
```

This example first defines a matrix (`sales_matrix`) that contains the sales for three products over four days. The key line is `daily_sales_growth = np.diff(sales_matrix, axis=0)`. By specifying `axis=0`, you tell NumPy to perform the difference operation along the rows (vertically). The output matrix shows the new sales each day for each product:

```
--- Daily Sales Growth ---
[[10  5 15]
 [15 10 20]
 [20 15 15]]
```

For instance, on the second day (the first row of the output), Product A sold 10 new units (110 – 100), Product B sold 5 (55 – 50), and Product C sold 15 (215 – 200).

Symbolic Calculus with SymPy

What if you don't have data, but a mathematical formula? For example, you might have a cost function $C(x) = 0.01x^2 + 50x + 5,000$. If you want to find the derivative or integral of the formula itself, you need a symbolic math library.

SymPy is the standard for this in Python. It works with abstract symbols, not numbers, to give you exact analytical results. Listing 5-5 uses SymPy for a cost function.

LISTING 5-5: USING SYMPY FOR A COST FUNCTION

```python
import sympy as sp

# 1. Define 'x' as a mathematical symbol
x = sp.symbols('x')

# 2. Create a symbolic expression for our cost function
cost_function = 0.01*x**2 + 50*x + 5000

# 3. Calculate the derivative (marginal cost function)
marginal_cost_function = sp.diff(cost_function, x)

# 4. Calculate the integral (if we wanted to reverse the process)
integral_of_cost = sp.integrate(cost_function, x)

print(f"Original Cost Function: {cost_function}")
print(f"Marginal Cost Function: {marginal_cost_function}")
print(f"Integral of Cost Function: {integral_of_cost}")
```

The code first defines x as an abstract symbol using `sp.symbols('x')`. This tells Python to treat x not as a variable holding a specific number, but as a mathematical placeholder, allowing you to build the `cost_function` as a symbolic expression.

Once you have this expression, you use SymPy's core calculus functions. `sp.diff(cost_function, x)` takes the expression and calculates its exact derivative with respect to x, giving the marginal cost function.

Conversely, `sp.integrate(cost_function, x)` finds the indefinite integral of the expression, effectively reversing the differentiation process.

The output confirms that SymPy has performed these symbolic manipulations correctly:

```
Original Cost Function: 0.01*x**2 + 50*x + 5000
Marginal Cost Function: 0.02*x + 50
Integral of Cost Function: 0.00333333333333333*x**3 + 25.0*x**2 + 5000.0*x
```

The code determined the derivative of the cost function, which represents the marginal cost. It also found the indefinite integral, reversing the operation. Notice that the integral includes a term, showing how it recovered the original structure from the constant term in your cost function.

Advanced Numerical Methods with SciPy

While NumPy provides the basics, SciPy provides a library of more advanced, high-performance numerical routines. When it comes to calculus, its `scipy.integrate` module is essential. You will see how to use `solve_ivp` later in this chapter to solve differential equations. Another key function is `quad`, which can find the definite integral of any Python function with high precision.

Let's say you have a Python function that represents your marginal revenue. You can use `scipy.integrate.quad` to find the total revenue gained by selling a certain range of products for instance, from the 100th unit to the 500th unit. Listing 5-6 shows an example of using the quad function.

LISTING 5-6: USING SCIPY TO INTEGRATE A FUNCTION

```python
from scipy import integrate

# Our marginal revenue function: MR = 100 - 0.1*q
def marginal_revenue(q):
    return 100 - 0.1 * q

# Integrate the function from quantity=100 to quantity=500
# The 'quad' function returns the result and an estimate of the error
total_revenue_gained, error = integrate.quad(marginal_revenue, 100, 500)

print(f"Total revenue from selling units 100 to 500: ${total_revenue_gained:.2f}")
print(f"Estimated error of the calculation: {error}")
```

The `integrate.quad()` function takes three main arguments: the function you want to integrate (`marginal_revenue`) and the start and end points of the integration interval (100 and 500). It returns two values: the calculated integral and an estimate of the absolute error.

The output from this listing is as follows:

```
Total revenue from selling units 100 to 500: $34000.00
Estimated error of the calculation: 3.774758283725532e-13
```

The extremely small error value indicates that the total revenue figure of $34,000.00 is highly precise. The quad function is the go-to tool when you need a precise numerical integral of a Python function.

Choosing the Right Tool

You can clearly see that Python can solve any problem. Which tool is the right one depends on your problem. The following can serve as a guide for what tool to use when:

➤   Starting with a spreadsheet or data array? Use NumPy.

➤   Starting with a mathematical formula? Use SymPy.

➤   Need to solve a differential equation or find a high-precision integral of a function? Use SciPy.

SOLVING BUSINESS GROWTH AND PRICING MODELS WITH DIFFERENTIAL EQUATIONS

While analyzing historical data is essential, the ultimate goal for many businesses is to look into the future. Businesses need to build models that can forecast growth, predict market saturation, and understand how systems evolve over time. This is the domain of differential equations.

If you think of the previous examples as reading a ship's log to see where it has been, think of differential equations as using the ship's current speed and the ocean currents to chart its future course. A differential equation is a powerful rule or "recipe" that describes how a system changes, where the rate of change depends on the current state of the system.

This concept is at the heart of many business dynamics:

➤   **Viral marketing:** The rate of new sign-ups is proportional to the number of existing users who can share the product. More users equals faster growth.

➤   **Compound interest:** The rate at which your investment grows is proportional to the current balance. More money equals faster earnings.

➤   **Product adoption:** The rate of new customers often depends on how many potential customers are *left* in the market. As the market becomes saturated, growth naturally slows down.

The example of product adoption is famously modeled by the logistic growth curve (or S-curve), which describes the lifecycle of many products and services: a slow start, a period of rapid growth, and a final leveling-off as the market capacity is reached.

To solve these models and generate forecasts in Python, you can use a powerful tool from the SciPy library: solve_ivp (Solve Initial Value Problem). You simply provide the solve_ivp function with a starting point (your initial number of users), a function that defines your "recipe for growth," and a time frame. It does the hard work of simulating the system's evolution.

As an example, consider the growth of a new streaming service. You have 100 beta testers at launch. Based on market research, your company believes the total potential market is 10,000 subscribers. You want to project subscriber growth over the next three years.

The "recipe for growth" is the logistic model, which states that the growth rate is proportional to both the current number of users and the remaining market space. Listing 5-7 contains code for modeling this.

LISTING 5-7: MODELING A SUBSCRIPTION SERVICE'S GROWTH

```python
import numpy as np
from scipy.integrate import solve_ivp
import matplotlib.pyplot as plt

# --- 1. Define the Model Parameters ---
P0 = 100      # Initial number of subscribers
K = 10000     # Market capacity (carrying capacity)
r = 1.5       # Intrinsic growth rate (e.g., 150% per year)

# --- 2. Define the Differential Equation (The "Recipe") ---
# This function describes the rate of change at any given time t and population P
def logistic_growth(t, P):
    # The logistic equation: dP/dt = r * P * (1 - P/K)
    return r * P * (1 - P / K)

# --- 3. Set up and Run the Solver ---
# Time span: 0 to 3 years
t_span = [0, 6]
# Points in time where we want the solution
t_eval = np.linspace(t_span[0], t_span[1], 100)

# Use solve_ivp to get the solution
sol = solve_ivp(
    fun=logistic_growth,
    t_span=t_span,
    y0=[P0],
    t_eval=t_eval
)
```

Let's break down how `solve_ivp` works. It's a sophisticated numerical solver that you set up by passing it four key arguments:

➤ fun, which is your recipe, the `logistic_growth` function you define to tell the solver how to calculate the rate of change at any given moment.

➤ t_span, the start and end times for your simulation, [0, 6] years.

➤ y0, the starting state of your system, which is the initial 100 subscribers ([P0]).

> ➤ t_eval, an optional but useful argument that tells the solver exactly which points in time you want it to report back.

This example asks for 100 evenly spaced points over the six years to ensure you get a smooth curve for plotting.

Under the hood, solve_ivp starts at P0 and takes a tiny step forward in time. It uses the function to calculate the growth rate, estimates the new subscriber count, and then repeats this process thousands of times. By stitching together these incremental steps, it accurately simulates the continuous growth curve without having to solve the complex differential equation by hand.

To better see what is happening, you can plot the results to see a graph of the results. Combine the code presented in Listing 5-8 with the code in Listing 5-7.

LISTING 5-8: VISUALIZING THE RESULTS OF THE SUBSCRIPTION SERVICE'S GROWTH

```
# --- 4. Visualize the Results ---
plt.figure(figsize=(10, 6))
plt.plot(sol.t, sol.y[0], label='Projected Subscriber Growth', color='purple')
plt.axhline(y=K, color='grey', linestyle='--', label=f'Market Capacity ({K})')
plt.axhline(y=K/2, color='red', linestyle=':', label='Point of Max Growth')

plt.title('Subscription Service Growth Projection (S-Curve)')
plt.xlabel('Years')
plt.ylabel('Number of Subscribers')
plt.legend()
plt.grid(True)
plt.show()
```

The output, as can be seen in Figure 5-4, is a classic S-curve that provides a rich, strategic forecast for the business. Looking at the figure, you can see the following:

> ➤ **Slow start (Year 0–2):** In the first year, growth is modest as the service finds its footing with early adopters.

> ➤ **Rapid growth (Year 2–4):** As word-of-mouth and marketing kick in, the service enters a period of exponential growth. The rate of new subscribers is at its highest when you have half the market (the dotted line), as there's a perfect balance of existing users to spread the word and new customers to acquire.

> ➤ **Saturation (Year 4–6):** As the service captures a majority of the potential market, growth slows significantly. It becomes harder and more expensive to find new subscribers.

This single model, born from a simple differential equation, is a powerful strategic roadmap. It helps you anticipate when to ramp up server capacity, when to shift marketing from acquisition to retention, and how to set realistic growth targets for investors.

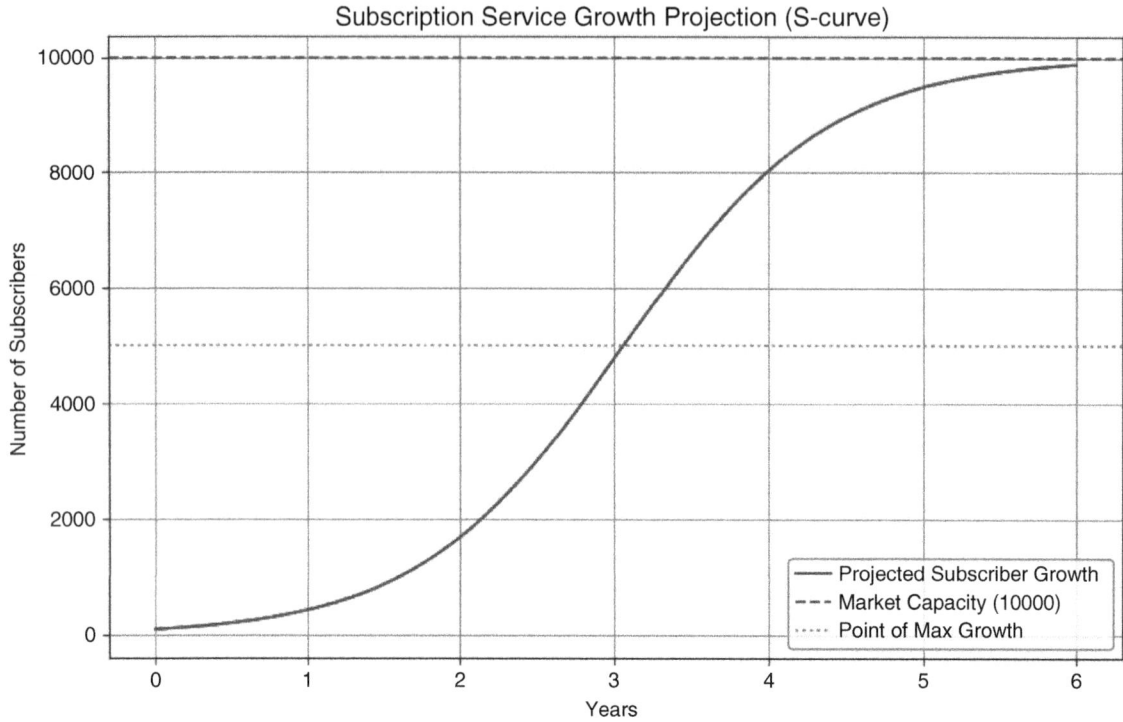

FIGURE 5-4: Subscription service growth project figure.

SENSITIVITY ANALYSIS WITH PARTIAL DERIVATIVES

In business, decisions are rarely made in a vacuum. A product's profit doesn't just depend on its price; it's also affected by your marketing budget, production costs, and competitor actions. This chapter has been looking at how one variable changes another, but the real world is a landscape of interconnected factors. Partial derivatives are the tool you use to navigate this landscape.

A partial derivative lets you measure the rate of change of a function with respect to one variable, while holding all other variables constant. It answers crucial "what-if" questions at the heart of business strategy:

➤   "If we keep our ad spend the same, how will a $1 price increase affect our profit?"

➤   "Holding our price steady, what is the exact return on the next dollar we put into marketing?"

This is like standing on a hillside—the slope depends on the direction you're facing. The partial derivative tells you the steepness if you decide to walk *only* east, or *only* north.

This next example models the profit for an online course. The profit depends on two key levers you control: the price of the course and your monthly marketing spend. Listing 5-9 presents a model of this and visualizes it as a 3D "profit surface" to see how these decisions interact.

LISTING 5-9: VISUALIZING A PROFIT LANDSCAPE

```python
import numpy as np
import matplotlib.pyplot as plt

# Define a profit function with two variables
def calculate_profit(price, marketing_spend):
    # Model how price affects quantity sold (higher price -> fewer sales)
    quantity_sold = 1000 - 10 * price + 2 * np.sqrt(marketing_spend)
    # Model revenue
    revenue = quantity_sold * price
    # Model cost
    cost = 2000 + (0.5 * marketing_spend) # Fixed cost + marketing
    return revenue - cost

# --- Create a Grid of Inputs ---
# Generate a range of possible prices and marketing spends
price_range = np.linspace(50, 150, 50)
marketing_range = np.linspace(500, 5000, 50)

# np.meshgrid creates a coordinate grid from our two arrays
P, M = np.meshgrid(price_range, marketing_range)

# --- Calculate Profit Across the Grid ---
Z_profit = calculate_profit(P, M)

# --- Visualize the Profit Surface ---
fig = plt.figure(figsize=(12, 8))
ax = fig.add_subplot(111, projection='3d')

surf = ax.plot_surface(P, M, Z_profit, cmap='viridis', edgecolor='none')

ax.set_title('Profit Landscape')
ax.set_xlabel('Price ($)')
ax.set_ylabel('Marketing Spend ($)')
ax.set_zlabel('Profit ($)')
fig.colorbar(surf, shrink=0.5, aspect=5, label='Profit')
plt.show()
```

Listing 5-9 first defines a `calculate_profit` function that takes `price` and `marketing_spend` as inputs. It models a realistic scenario where higher prices reduce sales quantity, while higher marketing spend increases it (but with diminishing returns, modeled by `np.sqrt`).

To visualize this, you need to calculate profit for many different combinations of price and marketing. `np.linspace` is used to create ranges of 50 evenly spaced values for both `price` and `marketing`. Then, `np.meshgrid` combines these two 1D arrays into a 2D grid of coordinates. This grid represents every possible combination of price and marketing spend to test.

You then pass these entire grids, `P` and `M`, to the profit function, which returns `Z_profit`, a matrix containing the profit for every point on that grid. Finally, you use Matplotlib's 3D plotting capabilities to render this as a surface. The 3D plot created from this listing is presented in Figure 5-5.

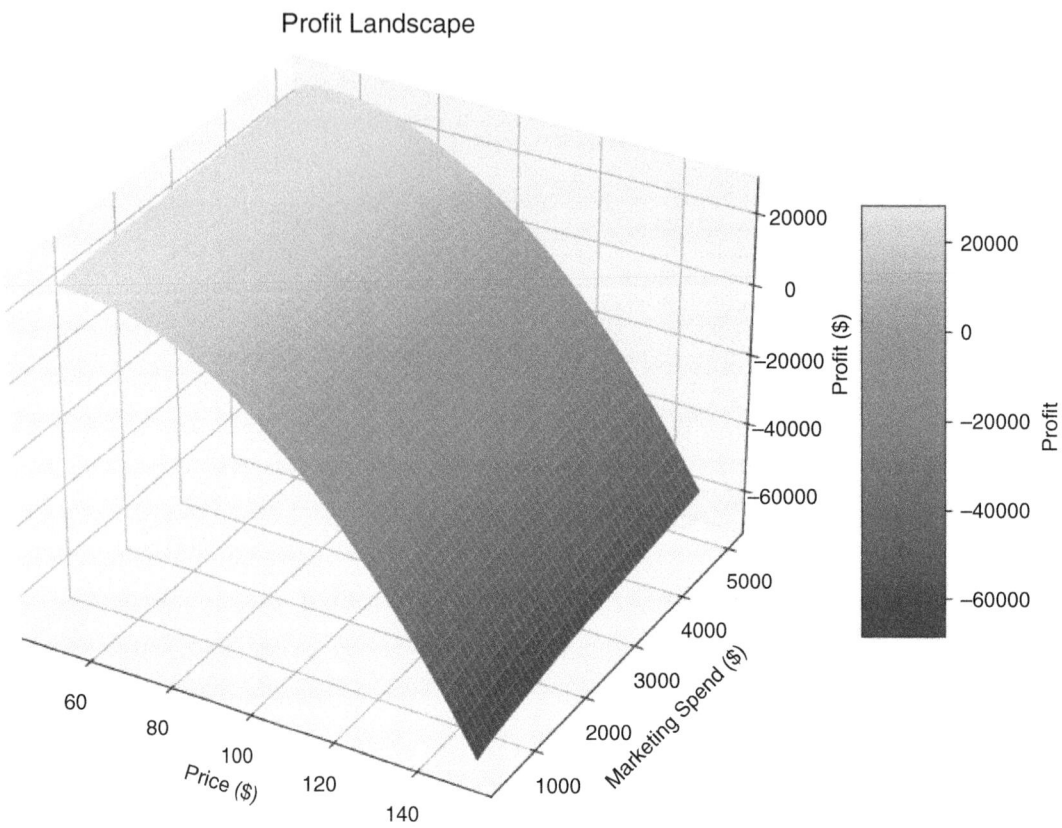

FIGURE 5-5: 3D plot of the profit landscape.

This plot shows an intuitive feel for the profit landscape. You can see there's a "ridge" where profit is maximized. If you set the price too high or too low, your profit falls off. Similarly, there are diminishing returns to marketing spend. This visualization turns a complex, two-variable problem into an intuitive map. A partial derivative at any point on this surface would tell you the slope in the direction of either the "Price" axis or the "Marketing Spend" axis, giving you a precise measurement of that variable's impact.

CASE STUDY: REVENUE, COST, AND PROFIT ANALYSIS

This case study brings everything together. You've seen how derivatives reveal rates of change and integrals accumulate totals. Now, you'll see how to apply these concepts to one of the most fundamental tasks in business: analyzing profitability.

In this case study, you'll act as an analyst for a company launching a new product. You have data on production costs and a model for market demand. The goal isn't to find the single "best" price (I save that for the next chapter on optimization), but to understand the dynamics of profit. Where does it grow fastest? When does producing one more unit stop being as profitable?

Step 1: Understanding Marginal Cost (the Derivative of Cost)

Your production team has given you a table of total costs for different production levels. As you produce more, you expect costs to rise, but due to economies of scale, they might not rise at a constant rate. The key metric you need to uncover is the marginal cost: the cost of producing one more unit. This is the derivative of the total cost. Using the code in Listing 5-10, you can uncover the margin cost.

LISTING 5-10: UNCOVERING THE MARGINAL COST

```python
import numpy as np
import matplotlib.pyplot as plt

# Production levels from 0 to 1000 units, in steps of 100
quantity = np.arange(0, 1001, 100)

# Total cost to produce at each level (includes fixed costs + variable costs)
# Let's model this with a function for simplicity
fixed_cost = 5000
variable_cost_per_unit = 50
# Add a non-linear term to represent economies of scale
total_cost = fixed_cost + (variable_cost_per_unit * quantity) -
(0.01 * quantity**2)

# Calculate Marginal Cost using the derivative (np.diff)
# We divide by the change in quantity (100) to get the per-unit cost
marginal_cost = np.diff(total_cost) / np.diff(quantity)

# --- Visualization ---
fig, (ax1, ax2) = plt.subplots(1, 2, figsize=(14, 6))

# Plot Total Cost
ax1.plot(quantity, total_cost, 'b-o', label='Total Cost')
ax1.set_title('Total Production Cost')
ax1.set_xlabel('Quantity Produced')
ax1.set_ylabel('Cost ($)')
ax1.grid(True)

# Plot Marginal Cost
# We plot this at the midpoint of each interval
ax2.plot(quantity[1:], marginal_cost, 'g-s', label='Marginal Cost')
ax2.set_title('Marginal Cost per Unit')
ax2.set_xlabel('Quantity Produced')
ax2.set_ylabel('Cost per Additional Unit ($)')
ax2.grid(True)

plt.tight_layout()
plt.show()
```

This listing first creates the data. `quantity` is an array representing production levels from 0 to 1,000 in steps of 100. You then model `total_cost` using a formula that includes a fixed cost, a variable

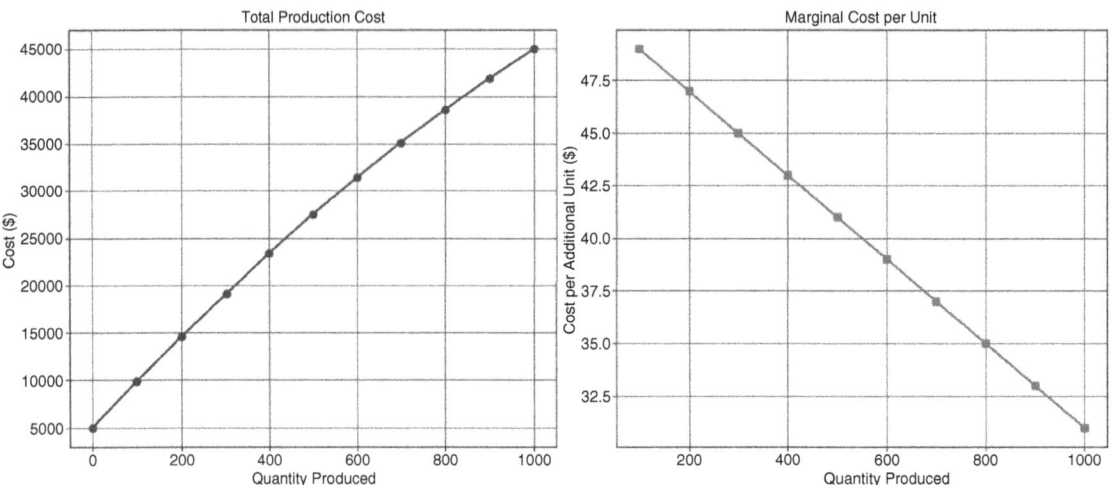

FIGURE 5-6: Total production cost and marginal cost per unit.

cost, and a small negative quadratic term (-0.01 * quantity**2) to simulate economies of scale, where costs rise a bit slower as quantity increases.

The crucial step is calculating the marginal_cost. The code uses np.diff(total_cost) to find the change in cost between each production level. However, since the production levels change by 100 units at a time (not 1), you must divide this by np.diff(quantity) (which is 100) to get the true per-unit marginal cost.

The visualization code then creates two side-by-side charts (using plt.subplots(1, 2)) to compare two views of this data, as seen in Figure 5-6.

Looking at Figure 5-6, you can see that the left chart shows the total cost, which always goes up. The right chart, however, provides a deeper insight. The marginal cost is decreasing. For the first 100 units, each additional item costs about $49 to make. By the time you're producing 1,000 units, the cost for just one more has dropped to $31. This is a classic sign of economies of scale: production is getting more efficient at higher volumes.

Step 2: Understanding Marginal Revenue (the Derivative of Revenue)

Next, your marketing team provides a demand curve. They estimate that to sell more units, you need to lower the price. Total revenue is simply price * quantity. But you're interested in the marginal revenue: the extra revenue you get from selling one more unit. The marginal revenue is computed in Listing 5-11.

LISTING 5-11: COMPUTING MARGINAL AND TOTAL REVENUE

```
# From our demand model, we know the price we must set to sell a certain quantity
price = 100 - (0.05 * quantity)

# Calculate Total Revenue
total_revenue = price * quantity

# Calculate Marginal Revenue using the derivative
marginal_revenue = np.diff(total_revenue) / np.diff(quantity)

# --- Visualization ---
fig, (ax1, ax2) = plt.subplots(1, 2, figsize=(14, 6))

# Plot Total Revenue
ax1.plot(quantity, total_revenue, 'r-o', label='Total Revenue')
ax1.set_title('Total Revenue')
ax1.set_xlabel('Quantity Sold')
ax1.set_ylabel('Revenue ($)')
ax1.grid(True)

# Plot Marginal Revenue
ax2.plot(quantity[1:], marginal_revenue, 'm-s', label='Marginal Revenue')
ax2.set_title('Marginal Revenue per Unit')
ax2.set_xlabel('Quantity Sold')
ax2.set_ylabel('Revenue from Additional Unit ($)')
ax2.grid(True)

plt.tight_layout()
plt.show()
```

This listing first defines the price based on the demand model: for every unit you want to sell, the price must drop by $0.05. It then calculates `total_revenue` by multiplying this `price` array by the `quantity` array.

Just as you did with cost, you calculate `marginal_revenue` by taking the derivative of total revenue using `np.diff(total_revenue)` and dividing it by the change in quantity (`np.diff(quantity)`).

The visualization code then generates two plots side-by-side, as shown in Figure 5-7.

As you can see in Figure 5-7, which shows the graphs that are output from the listing, the total revenue curve on the left has an interesting shape. It peaks and then starts to decline. The marginal revenue chart on the right shows why. Initially, selling one more unit brings in about $95. But as you produce more, you have to lower the price on all units, so the extra revenue you get from each new sale decreases. Eventually, the marginal revenue becomes negative. This means that to sell one more unit, the price drop required is so large that your total revenue actually goes down.

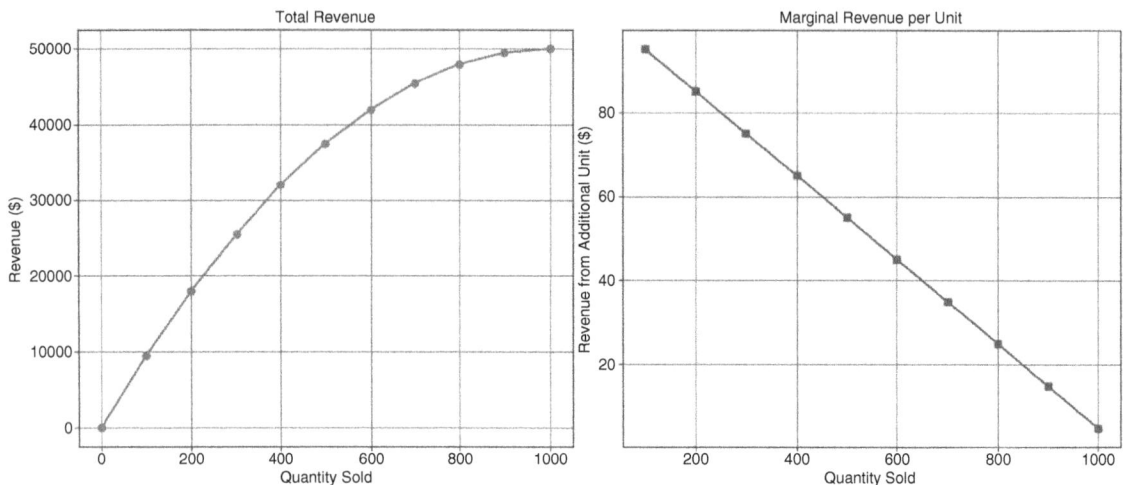

FIGURE 5-7: Total revenue and marginal revenue per unit.

Step 3: Finding the Sweet Spot with Marginal Profit

Now you need to combine the two. Profit is revenue minus cost. The most important analytical concept here is *marginal profit*, which is marginal revenue minus marginal cost. As long as the marginal profit is positive, producing one more unit adds to your total profit. The moment it turns negative, you start losing money on each new unit. Combine Listing 5-10, Listing 5-11 to calculate the profit margin for these products.

LISTING 5-12: CALCULATING MARGINAL PROFIT

```
# Calculate Total Profit
total_profit = total_revenue - total_cost

# Calculate Marginal Profit
marginal_profit = marginal_revenue - marginal_cost

# --- Final Visualization ---
plt.figure(figsize=(12, 7))

plt.plot(quantity[1:], marginal_revenue, 'm--', label='Marginal Revenue')
plt.plot(quantity[1:], marginal_cost, 'g--', label='Marginal Cost')
plt.plot(quantity[1:], marginal_profit, 'k-o', label='Marginal Profit (MR - MC)')
plt.axhline(y=0, color='grey', linestyle='-')

plt.title('Marginal Analysis of Profitability')
plt.xlabel('Quantity')
plt.ylabel('Dollars per Unit ($)')
plt.legend()
plt.grid(True)
plt.show()
```

This code performs a simple element-wise subtraction on the NumPy arrays to calculate `total_profit` and `marginal_profit`. The visualization then brings all three key marginal metrics onto a single chart for comparison. Distinct styles are used for each line—dashed for revenue, lighter dashed line for cost, and a solid black line with markers for profit—to make them easy to distinguish. Crucially, a solid horizontal line is added at zero (`plt.axhline(y=0)`). This acts as the breakeven line for marginal profit, instantly showing you where adding another unit stops being profitable.

The plot shown in Figure 5-8 reveals the entire story. The solid downward-sloping line, marginal profit, shows the profit from each additional unit.

➤ **From 100 to 700 units:** The marginal profit is positive. Each new unit you produce and sell adds to your bottom line.

➤ **Around 700 units:** The marginal revenue and marginal cost curves intersect. At this point, the marginal profit is zero. You've squeezed all the profit you can out of the next unit.

➤ **Beyond 700 units:** The marginal profit becomes negative. The cost to produce one more unit is now higher than the revenue it brings in. You are losing money on every additional sale.

This analysis, driven entirely by the concept of the derivative, has shown that the most profitable production level is right around 700 units. While a formal optimization algorithm could pinpoint the exact number, this calculus-based analysis provides a powerful, visual understanding of the dynamics at play.

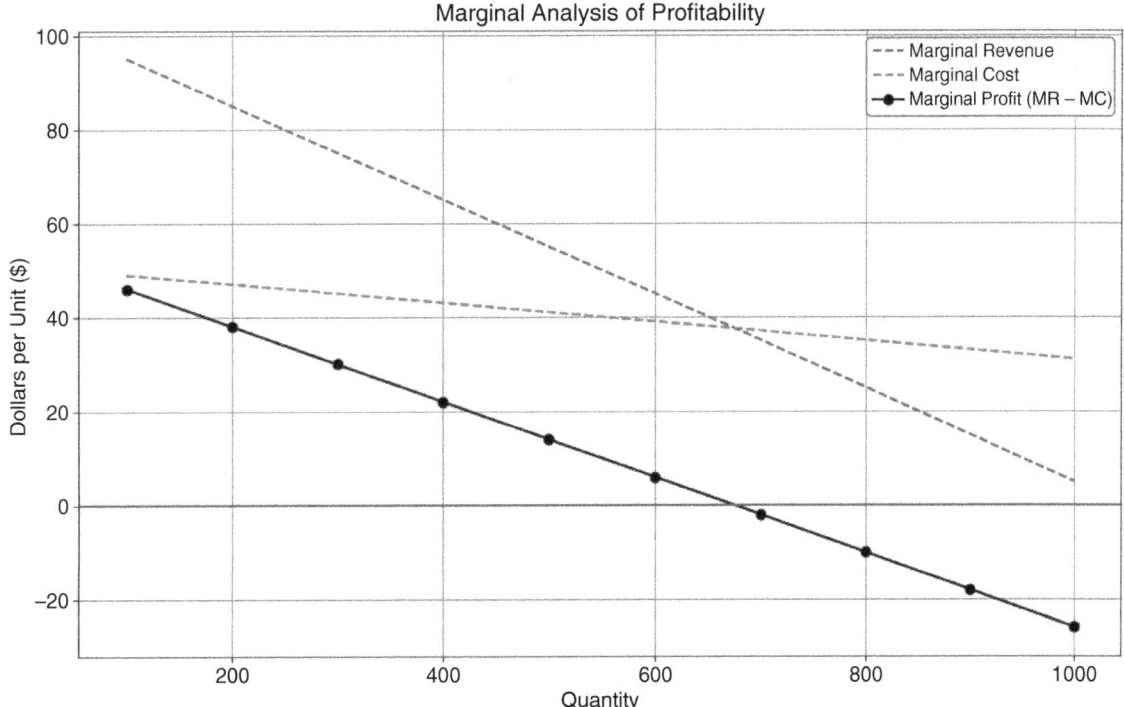

FIGURE 5-8: Marginal analysis of profitability line plot.

SUMMARY

Throughout this chapter, you have journeyed through the core concepts of calculus, transforming them from abstract mathematical ideas into practical tools for business analysis. The chapter began by using numerical derivatives and integrals to read the story hidden in data, uncovering the momentum of a product launch and the cumulative growth of a subscriber base. From there, you moved from analyzing the past to predicting the future, using differential equations to model the classic S-curve of product adoption. The chapter then embraced real-world complexity by using partial derivatives to explore the multi-dimensional "profit landscape," showing how different strategic levers interact.

Finally, the case study on marginal analysis tied everything together, showing how the derivative can reveal the precise dynamics of profitability. That analysis led to a powerful conclusion: the sweet spot for production was *around 700 units*.

This insight is also a question. *Around 700* is a fantastic analytical finding, but as business leaders, we need to make a specific decision. How do we find the *exact* point of maximum profit? To answer that, we must move from analyzing the landscape to finding its highest peak. That is the world of optimization, and it is where your journey takes you next.

CONTINUE YOUR LEARNING

If you want to dive deeper into the tools and concepts discussed in this chapter, the following official resources are highly recommended:

➤ NumPy documentation:

    ➤ The foundation of numerical computing in Python. The official documentation is the best place to learn more about the functions used in this chapter and discover new ones.

    ➤ **numpy.diff:** `https://numpy.org/doc/stable/reference/generated/numpy.diff.html`

    For understanding how to calculate differences between array elements.

    ➤ **numpy.cumsum:** `https://numpy.org/doc/stable/reference/generated/numpy.cumsum.html`

    For learning more about cumulative sums.

➤ SciPy documentation:

    ➤ SciPy builds on NumPy to provide a large collection of algorithms for scientific and technical computing.

    ➤ **scipy.integrate.solve_ivp:** `https://docs.scipy.org/doc/scipy/reference/generated/scipy.integrate.solve_ivp.html`

    The definitive guide to the initial value problem solver used for the differential equation modeling.

➤ Matplotlib documentation:

  ➤ The primary library for creating static, animated, and interactive visualizations in Python.

  ➤ **Official tutorials:** `https://matplotlib.org/stable/tutorials/index.html`

  A great starting point for mastering plotting, from basic charts to complex 3D surfaces.

Optimization Techniques for Business Strategy

In the last chapter, analyzing the marginal cost and revenue led to a powerful insight: the sweet spot for the product's profitability was around 700 units. The power of calculus gives you a deep, intuitive understanding of the dynamics at play. But as a business leader, "around 700" isn't a final decision. It's a fantastic analytical finding, but it leaves you with one critical, actionable question: what is the exact number?

This chapter focuses on answering that question. You are now moving from analysis, understanding the landscape, to prescription. This is the world of optimization, the formal process of finding the best possible outcome from a set of choices.

Optimization is the engine behind countless strategic decisions, from an airline setting its ticket prices to a CPG company deciding how to stock a supermarket's shelves. The good news is that every optimization problem, no matter how complex, is built from two simple pillars:

➤ **Objective function:** This is the single number you want to maximize or minimize. It's your goal, defined in code. It could be maximize profit, minimize cost, or minimize portfolio risk.

➤ **Constraints:** These are the real-world rules, limitations, and budgets you must operate within. Examples include "We only have 40 labor hours per week," "Our total marketing budget is $10,000," or "All portfolio weights must sum to 1."

This chapter shows you how to build a complete toolkit for making these optimal decisions using Python. You'll start by solving the very problem you ended with, and then build up to solving complex, real-world problems in logistics, finance, and marketing.

THE PYTHON OPTIMIZATION ECOSYSTEM

Before you jump into solving problems, it's helpful to know what tools are available. When you face an optimization problem, you're rarely starting from scratch. Your choice of tool depends on the nature of your problem. The Python ecosystem has a mature and powerful set of libraries designed to do the heavy lifting for you:

1. `scipy.optimize`—This is the workhorse module used throughout this chapter. It's part of the SciPy library and provides a fantastic toolkit for a wide range of optimization tasks. It's the perfect place to start because it contains tools that can handle almost any optimization problem:

   ➤ `minimize_scalar`: The tool you'll use for simple problems when you're optimizing just *one* variable (like finding the optimal quantity of a single product).

   ➤ `linprog`: A high-performance solver specifically designed for Linear Programming (LP) problems, like the kind of allocation problems that form the bedrock of operations and logistics.

   ➤ `minimize`: The most powerful and general-purpose tool in the box. This is what you use for complex nonlinear constrained optimization, like building an optimal financial portfolio.

2. **PuLP and Pyomo**—For very large-scale, complex linear or integer programming (where your variables must be whole numbers, like "5 trucks" or "27 employees"), specialized libraries like PuLP and Pyomo shine. Their main advantage is that they allow you to write your models in a more "algebraic" and human-readable way, which can be a lifesaver when you're managing hundreds of constraints for a factory floor or supply chain. You won't use these in this introductory chapter, but it's important to know they exist. You can find more information about PuLP at `https://coin-or.github.io/pulp/` or about Pyomo at `https://www.pyomo.org/documentation`.

3. **CVXPY (convex optimization)**—This is a more advanced library for a special class of problems known as *convex optimization*. These problems are common in finance, machine learning, and engineering, and they have a wonderful property: they are guaranteed to have a single, global "best" solution. CVXPY provides a powerful, high-level language to formulate and solve these specific types of problems. You can find more information about CVXPY in the "Continue Your Learning" section at the end of this chapter.

For the purposes here, `scipy.optimize` is the perfect tool. It's powerful, it's integrated directly with NumPy, and its functions map perfectly to the core concepts you need to learn—unconstrained, linear constrained, and nonlinear constrained optimization.

A FRAMEWORK FOR SOLVING MOST OPTIMIZATION PROBLEMS

Knowing the tools is the first step. The second, more difficult step is learning how to think like an optimizer. The hardest part of optimizing business models is not the Python code, but the art of formulation—translating a messy, real-world business problem into the clean, mathematical structure that a solver can understand.

All of the examples in this chapter use a simple, four-step framework.

The Four-step Formulation Process

Step 1. Identify Your Objective Function This is the most critical step. You must be able to state your goal as a *single function*—is it minimize cost, maximize profit, or minimize risk? Then, ask if the function is *linear* (a straight line, like Profit = 50*T + 30*C) or *nonlinear* (a curve, like the `total_profit` function from Chapter 5).

Step 2. Identify Your Constraints What are the rules? What limits your resources? This includes budgets, time, labor hours, or production capacity. Write them down as mathematical expressions. Are they equalities (=) or inequalities (<=, >=)?

Step 3. Choose Your Solver This is now a simple decision tree based on your answers to the questions in Step 2:

➤ One nonlinear variable and no constraints?

   Use `scipy.optimize.minimize_scalar`.

➤ A linear objective and linear constraints?

   Use `scipy.optimize.linprog`.

➤ A nonlinear objective with constraints?

   Use `scipy.optimize.minimize`.

Step 4. Formulate and Solve Translate your objective and constraints into the specific format that your chosen solver requires.

This is the "translation" step, where you convert your business rules into Python objects. For linear problems, you will format your objective as a coefficient list `c` and your constraints as matrices `A_ub` (inequality) and `b_ub` (bounds). For nonlinear problems, you will define a Python function called `fun` for your objective and a list of dictionaries for your constraints. Once formatted, you simply pass these arguments to the solver, run the code, and interpret the results.

Understanding the Local vs. Global Optima Issue

The most important trap with working with this framework centers on local versus global optima. You'll notice that some solvers need a starting or initial guess (which will call x0) and others don't. This is because of the single most important concept in nonlinear optimization: the difference between a local and a global optimum.

Linear problems like the product mix problem are simple in one sense. They are convex, meaning they have no hills or valleys, just a single solution. A solver like `linprog` is guaranteed to find this single global optimum every time, no matter where it starts.

Nonlinear problems like the profit curve can be non-convex. They can have multiple hills and valleys, like a rugged mountain range. Think of a curvy line on a plot.

These hills and valleys are called:

➤ A *local minimum,* which is a valley that is lower than all the points immediately around it.

➤ A *global minimum,* which is the absolute deepest valley in the entire landscape.

A solver (like `scipy.optimize.minimize`) can get stuck in a local minimum. Meaning, when following that valley, it can think it has reached the lowest point, when really the lowest point is over the next hill. The solver will find the bottom and report back that it's found a solution, because as far as it can "see," there's no lower point nearby.

This is why the initial guess (x0) can be important. Starting the search at a different point on the mountain might lead the solver into a different, and hopefully deeper, valley. This is also why `minimize_scalar`'s bounds argument is so useful; it tells the solver to only search for the lowest valley within a specific, bounded section of the landscape.

Applying the Framework: Profit Maximization

With a framework in place, let's re-examine the first problem from Chapter 5. Let's start by answering the question from Chapter 5: what is the *precise* quantity that maximizes profit?

This is the simplest form of optimization, known as *unconstrained nonlinear optimization.* This is *nonlinear* because the profit curve (driven by revenue and cost models) is a curve, not a straight line. It is *unconstrained* because for this first example, you can assume there are no constraints. You can produce any number of units you want.

The tool for this job is `scipy.optimize.minimize_scalar`. This function is a hill-climbing (or, more accurately, valley-finding) algorithm. You give it a function and a starting guess, and it intelligently searches for the single lowest point on that function.

To find a maximum (like profit), you just have to be a little clever: you ask the function to find the minimum of negative profit. By putting a negative sign in front of profit, you can use the minimize function to find the lowest point, which will actually be the highest (thanks to the negative). I call this the negative profit trick and it works for nearly any value.

First, let's rebuild the `total_profit` function from the Chapter 5 case study. This is done in Listing 6-1, where functions are defined for revenue and cost, and then combined into a single `total_profit` function.

LISTING 6-1: PINPOINTING MAXIMUM PROFIT

```python
import numpy as np
import matplotlib.pyplot as plt
from scipy.optimize import minimize_scalar

# --- Step 1: Re-create the functions from Chapter 5 ---

def calculate_total_revenue(quantity):
    """Calculates total revenue based on the demand curve."""
    price = 100 - (0.05 * quantity)
    return price * quantity

def calculate_total_cost(quantity):
    """Calculates total cost based on fixed and variable costs."""
    fixed_cost = 5000
    variable_cost_per_unit = 50
    # Add nonlinear term for economies of scale
    return fixed_cost + (variable_cost_per_unit * quantity) - (0.01 * quantity**2)

def calculate_total_profit(quantity):
    """Calculates total profit (Revenue - Cost)."""
    return calculate_total_revenue(quantity) - calculate_total_cost(quantity)

# --- Step 2: Create the Objective Function ---
# Our optimizer *minimizes*. To *maximize* profit, we
# create an objective function that returns the *negative* profit.
def objective_function(quantity):
    return -calculate_total_profit(quantity)

# --- Step 3: Run the Optimizer ---
# 'minimize_scalar' finds the minimum of our objective_function.
# We give it a 'bracket' (a range [0, 2000]) to search within.
result = minimize_scalar(objective_function, bounds=(0, 2000), method='bounded')

# The optimal quantity is stored in 'result.x'
optimal_quantity = result.x
max_profit = calculate_total_profit(optimal_quantity)

print(f"--- Optimization Result ---")
print(f"Optimal Production Quantity: {optimal_quantity:.2f} units")
print(f"Maximum Profit Achieved: ${max_profit:.2f}")

# --- Step 4: Visualize the Result ---
quantity_range = np.linspace(0, 1500, 400)
profit_range = calculate_total_profit(quantity_range)

plt.figure(figsize=(10, 6))
plt.plot(quantity_range, profit_range, label='Total Profit')
plt.axvline(x=optimal_quantity, color='red', linestyle='--',
            label=f'Optimal Point ({optimal_quantity:.0f} units)')
plt.title('Finding the Exact Point of Maximum Profit')
plt.xlabel('Quantity Produced')
plt.ylabel('Total Profit ($)')
```

```
plt.legend()
plt.grid(True)
plt.show()
```

Step 1 essentially translates the business logic into Python functions. `calculate_total_revenue` takes a quantity and returns the revenue based on the linear demand curve. `calculate_total_cost` does the same for the cost model, including the nonlinear term for economies of scale. `calculate_total_profit` is simply the difference between the two. These are the fundamental building blocks of this model.

Step 2 adapts this model for the optimizer. A new function, `objective_function`, takes one input (quantity) and returns one output: the negative profit. This is the function you will pass to `minimize_scalar`.

Step 3 calls the solver. You can use `minimize_scalar` because you only have one variable to change (quantity). You pass it to the `objective_function`, and critically, also pass a `bounds=(0, 2000)` argument and set `method='bounded'`. This tells the solver not to waste time looking for negative production quantities or unrealistically high ones. It focuses the search on the relevant range.

The solver returns a result object. The most important attribute is `result.x`, which holds the optimal input value it found, in this case, the optimal production quantity.

The results of the optimization are printed to the screen:

```
--- Optimization Result ---
Optimal Production Quantity: 714.29 units
Maximum Profit Achieved: $10714.29
```

Finally, Step 4 creates a visualization to confirm the results, which is shown in Figure 6-1. This step generates a range of quantity values, calculates the profit for each, and plots the curve. It then uses `plt.axvline` to draw a red dashed line at the `optimal_quantity` found by the solver.

As Figure 6-1 shows, the dashed line perfectly intersects the peak of the profit curve. You have successfully moved from knowing the peak was "around 700" to knowing it is exactly at 714.29 units.

LINEAR PROGRAMMING

This section turns your attention to one of the most powerful and widely used optimization techniques in the entire business world: *linear programming* (LP).

This is the quintessential tool for solving problems of allocation and logistics. Linear programming is used to determine the most efficient way to use limited resources. It's the magic behind airline scheduling, factory production planning, and supply chain management.

A problem is linear if all of its relationships are straight lines:

- ➤ **Linear objective:** Each additional unit of a product adds a fixed amount of profit (e.g., $10 profit per chair).

- ➤ **Linear constraints:** Each additional unit consumes a fixed amount of a resource (e.g., 2.5 hours of labor per chair).

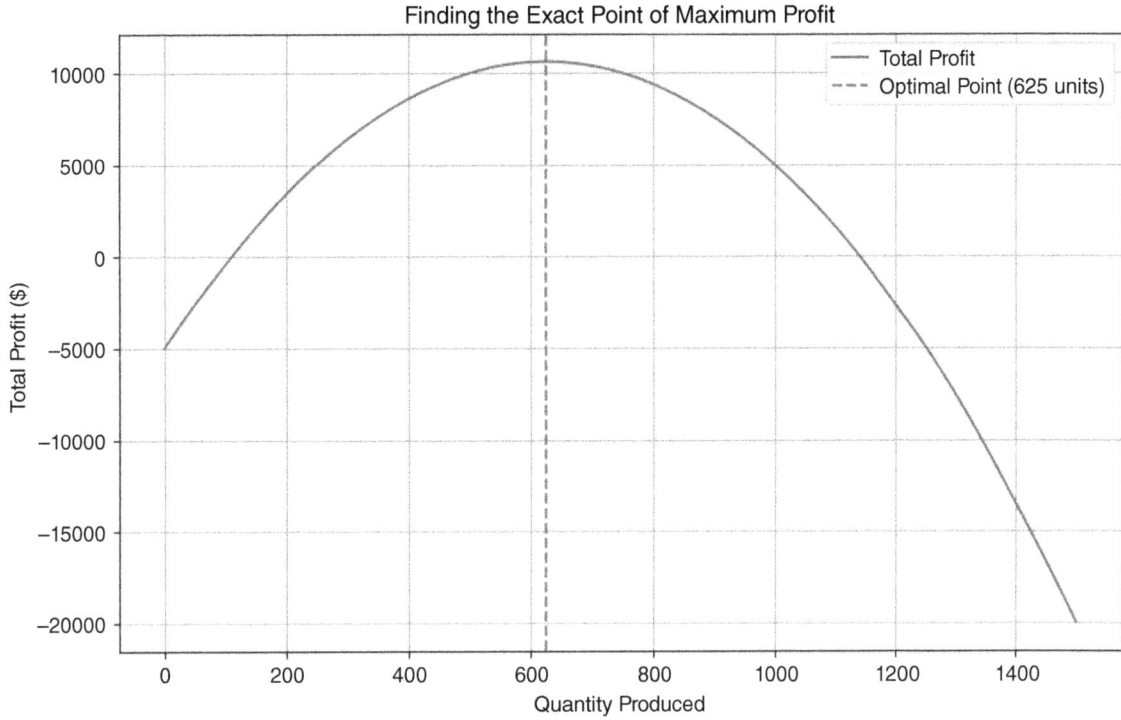

FIGURE 6-1: Visualization of the maximum profit.

This introduces the second core concept: *constrained optimization.* Unlike with the first example, you now have to operate within a set of rules. You can't just make infinite products; you're limited by certain constraints (e.g., budget, time, raw materials). Without these boundaries, a non-constant linear objective function would simply continue indefinitely, meaning there would be no reachable maximum or minimum value to find.

The tool for this is `scipy.optimize.linprog`. This function is specifically designed to solve these linear problems at high speed.

Consider an example where you own a small workshop that makes two products:

➤ **Tables:** A table sells for a profit of $50.

➤ **Chairs:** A chair sells for a profit of $30.

You have two main constraints:

➤ **Labor:** You have a total of 120 labor hours available per week.

    ➤ Each table requires six hours of labor.

    ➤ Each chair requires three hours of labor.

➤ **Materials:** You have a total of 200 units of wood available per week.

    ➤ Each table requires eight units of wood.

    ➤ Each chair requires five units of wood.

What is the exact number of tables and chairs you should produce to maximize your total profit, without exceeding the resource constraints? Before you can write any code, you must translate this business problem into the formal language of optimization. This is the most critical step.

Step 1. Objective function: This is what you want to maximize. Let's use T for the number of tables and C for the number of chairs. The total profit is the sum of the profit from tables and chairs. The function to use in order to make your profit as large as possible is as follows:

```
Profit = 50*T + 30*C
```

Step 2. Constraints: These are the rules of the workshop, the real-world limits you cannot break. Your constraints for this problem are as follows:

➤ **Labor constraint:** The total labor used must be less than or equal to 120 hours.

```
6*T + 3*C <= 120
```

➤ **Wood constraint:** The total wood used must be less than or equal to 200 units.

```
8*T + 5*C <= 200
```

➤ **Non-negativity constraint:** You can't make a negative number of products.

```
T >= 0 and C >= 0
```

This complete formulation is your "blueprint." Now you can prepare these components for the `scipy.optimize.linprog` function.

You can use `linprog` in Listing 6-2 to solve this. One important detail: `linprog`, like the previous tool, is a minimizer by default. To make it maximize the profit, you simply provide it with the negative of your profit coefficients (–$50, –$30). Minimizing this negative value is mathematically the same as maximizing a positive one.

LISTING 6-2: THE PRODUCT MIX PROBLEM

```
import numpy as np
from scipy.optimize import linprog

# --- Step 1: Define the Problem ---

# Objective function (to be minimized):
# We want to MAXIMIZE: 50*T + 30*C
# So we MINIMIZE:     -50*T - 30*C
c = [-50, -30]  # Coefficients for T (Table) and C (Chair)

# Constraints (Left-hand side):
# Ax_b <= b_b
# Constraint 1 (Labor): 6*T + 3*C <= 120
```

```
# Constraint 2 (Wood):  8*T + 5*C <= 200
A = [
    [6, 3],  # Labor constraints
    [8, 5]   # Wood constraints
]

# Constraints (Right-hand side):
b = [
    120, # Max labor hours
    200  # Max units of wood
]

# --- Step 2: Define the Bounds ---
# We can't make negative products
T_bounds = (0, None)  # (min_tables, max_tables)
C_bounds = (0, None)  # (min_chairs, max_chairs)

# --- Step 3: Run the Optimizer ---
result = linprog(c, A_ub=A, b_ub=b, bounds=[T_bounds, C_bounds], method='highs')

# --- Step 4: Interpret the Results ---
optimal_tables = result.x[0]
optimal_chairs = result.x[1]
# Multiply by -1 to convert minimized negative profit back to max profit
max_profit = -result.fun

print("--- Optimal Production Plan ---")
print(f"Optimal number of Tables to produce: {optimal_tables:.0f}")
print(f"Optimal number of Chairs to produce: {optimal_chairs:.0f}")
print(f"Maximum Profit Achieved: ${max_profit:.2f}")
```

Listing 6-2 translates the blueprint into Python lists. The objective function coefficients go into list c. The constraints are split into two parts: the coefficients for the variables go into matrix A (the left side), and the limits go into list b (the right side). You then need to define bounds for each variable to ensure they are non-negative.

Finally, the code calls linprog, passing it all these pieces. The method='highs' argument tells SciPy to use its most modern and efficient linear programming solver. The beauty of this setup is its flexibility. If you wanted to add a third product, such as a desk, you would simply add a third coefficient to c, a third column to A, and a third set of bounds. If you added a new constraint, such as an allotment for metal, you would just add a new row to A and a new value to b.

```
The result of our optimization is:--- Optimal Production Plan ---
Optimal number of Tables to produce: 0
Optimal number of Chairs to produce: 40
Maximum Profit Achieved: $1200.00
```

This is a fascinating and non-obvious result! A quick glance might suggest that you should make tables since they have a higher profit margin. However, the optimizer shows that tables are too resource-intensive. They consume a large amount of your most limited resource: labor.

The optimizer's solution is to focus all the production on chairs. By making 40 chairs, you use all 200 units of wood (40 × 5) and all 120 hours of labor (40 × 3), maximizing the profit at $1,200. This is the power of linear programming: it finds the truly optimal solution within real-world constraints, which is often counterintuitive.

Constrained Optimization

So far within this chapter, you have seen the two fundamental types of optimization: unconstrained and constrained. Understanding the difference between these is the most important part of framing a business problem. Let's look at a summary of each.

Unconstrained Optimization: Finding the True Peak

Finding the true peak is the problem you solved in Listing 6-1. You had a single, nonlinear profit curve, and you wanted to find the absolute highest point on it. Let's look at the details of this problem:

➤ **Analogy:** Finding the highest peak in an entire mountain range.

➤ **Business question:** "What is the absolute best price for our product to maximize profit, assuming we can make as many as we need and there are no other limits?"

➤ **The tool:** `scipy.optimize.minimize_scalar` was perfect for this. It's designed to just search until it finds the peak.

Constrained Optimization: Finding the Best Possible Point

Finding the best possible point is the problem you solved in Listing 6-2. You had a profit "surface" (defined by tables and chairs), but were fenced in by real-world limits (labor and wood). Let's again look at the details of this problem:

➤ **Analogy:** Find the highest point, but don't leave the boundaries of a fenced-in park. The highest point in the park might just be a small hill, not the highest peak in the whole mountain range.

➤ **Business question:** "What is the best product mix we can create given our limited $50,000 budget, 200 weekly labor hours, and 1,000 units of raw material?"

➤ **The tool:** `scipy.optimize.linprog` is designed to find the best possible solution within the constrained area, which is known as the feasible region.

Almost every real-world business problem is a constrained problem. You never have infinite money, infinite time, or infinite resources. The key to solving them is to correctly identify and define your constraints.

The Geometry of Optimization

Before you move on to complex real-world applications, it is valuable to pause and visualize exactly what these algorithms are doing. Optimization is essentially a search for the highest (or lowest) point on a surface. The shape of that surface dictates which tool you use and whether you will succeed.

The first example optimized a simple profit curve. It had one clear peak. This is an ideal scenario known as a convex problem. But many real-world nonlinear problems are rugged. They have multiple hills and valleys.

Imagine a profit landscape where changing your price slightly moves you up a small hill, but a drastic price change might move you to a completely different, higher mountain range. A "hill-climbing" algorithm is essentially blind; it only knows the slope of the ground immediately beneath its feet. If you start it at the bottom of the small hill, it will climb to the top, look around, see that every direction is down, and incorrectly report that it has found the global maximum.

Listing 6-3 generates a 3D wireframe plot of such a rugged landscape to illustrate this danger.

LISTING 6-3: VISUALIZING A RUGGED OPTIMIZATION LANDSCAPE

```python
import numpy as np
import matplotlib.pyplot as plt

# 1. Define a "Rugged" Function
# This combines a large underlying curve with smaller sine wave ripples
def rugged_profit(x, y):
    base_structure = -(x**2 + y**2) # The main hill
    ripples = 10 * (np.cos(x) + np.sin(y)) # The local peaks/valleys
    return base_structure + ripples

# 2. Create the Grid
x = np.linspace(-5, 5, 50)
y = np.linspace(-5, 5, 50)
X, Y = np.meshgrid(x, y)
Z = rugged_profit(X, Y)

# 3. Visualize with Wireframe (Grayscale friendly)
fig = plt.figure(figsize=(10, 8))
ax = fig.add_subplot(111, projection='3d')

# Plot wireframe
ax.plot_wireframe(X, Y, Z, rstride=2, cstride=2, color='black', linewidth=0.5)

# Highlight Global Max vs Local Max
ax.scatter(0, 0, 20, color='black', s=100, label='Global Maxima (Goal)')
ax.scatter(3, 3, -10, color='gray', s=100, label='Local Maxima (Trap)')

ax.set_title('The Trap of Nonlinear Optimization: Local vs. Global')
ax.set_xlabel('Variable A')
ax.set_ylabel('Variable B')
ax.set_zlabel('Profit')
ax.legend()

plt.show()
```

Listing 6-3 intentionally constructs a difficult landscape. It starts with a `base_structure`, which is a simple inverted parabola (a smooth hill). Then, ripples are added using sine and cosine functions. This creates a surface that has a general shape but is covered in bumps and dips. `np.meshgrid` creates a grid of coordinates and calculates the height Z for every point. Finally, `ax.plot_wireframe` draws lines connecting these points, creating a net-like visual that clearly shows the terrain without needing color gradients. I manually added two scatter points to highlight the key difference: a local maxima (a small peak) and the global maxima (the true highest point). Figure 6-2 visualizes this optimization problem.

If the solver starts near the gray dot (the local maxima), it will climb to that peak and stop. It has no way of seeing the much higher, black dot (the global maxima) across the valley. This geometric reality is why choosing a good initial guess is critical in nonlinear optimization. It places the blind climber on the right mountain.

Contrast this with linear programming. A linear objective function (like Profit = 50*T + 30*C) is fundamentally different. It has no curves, no hills, and no valleys. It is a perfectly flat plane, tilted upward.

Because it is flat, a linear function has no critical points (peaks or valleys) in the middle. If you are standing on a tilted ramp, the "highest point" is never where you are standing; it is always farther up the ramp.

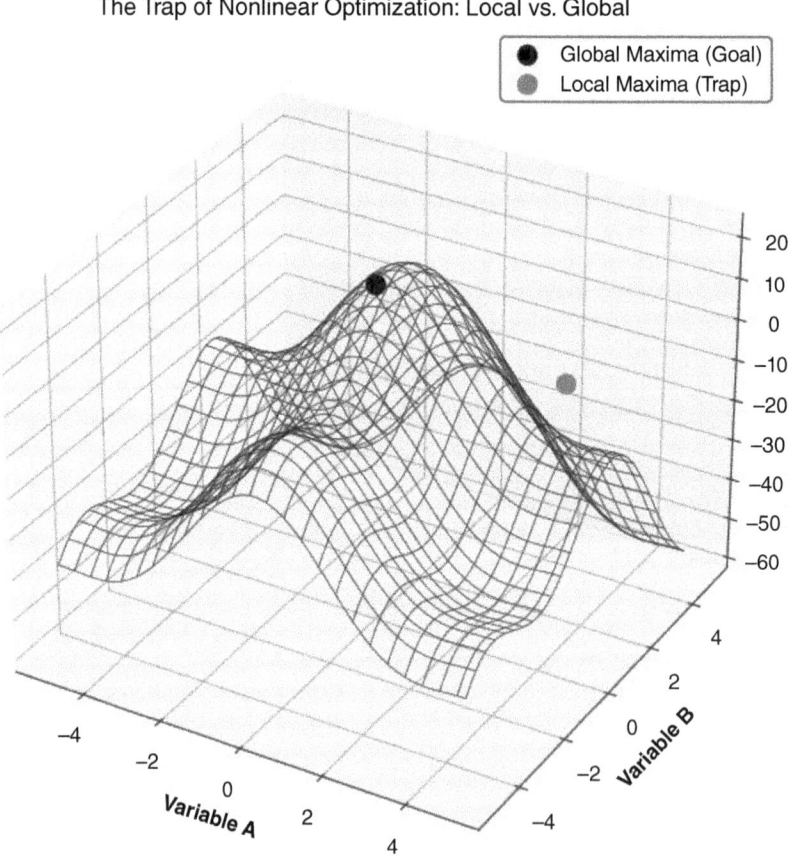

FIGURE 6-2: Local vs. global optimization.

This leads to two critical insights about linear programming:

➤ **Unboundedness:** Without constraints, a linear problem has no solution. The plane simply goes up forever toward infinity.

➤ **The importance of vertices:** When you add constraints (walls) to this infinite ramp, you create a geometric shape called a *polytope* (a multi-dimensional polygon). The optimal solution is guaranteed to lie at one of the vertices (or the corners) of this shape.

Imagine tilting a box. The lowest point will always be one of the corners. This is why the product mix solver returned exactly 40 chairs and 0 tables. It didn't suggest a blend like 38 chairs and 1.2 tables because the mathematical peak of a linear function can only exist at the extreme boundaries of the feasible region.

With more knowledge of the underlying algorithms, you can understand when they fail. Even the best solvers can fail. Understanding the geometry helps you diagnose why an optimization run crashes or returns nonsense.

➤ **Unboundedness:** As discussed, if you forget a constraint (e.g., you forget to limit the labor hours), a linear solver will realize it can make infinite profit. It will return an "Unbounded" error status, which essentially means: "You haven't defined the limits of reality yet."

➤ **Infeasibility:** This happens when your rules contradict each other. Imagine telling a solver, "Production must be greater than 100 units" and "Production must be fewer than 50 units." Geometrically, the "feasible region" (the area where all rules are met) does not exist. It is empty. The solver will return an "Infeasible" status, telling you that your business requirements are impossible to satisfy simultaneously.

➤ **No unique solution:** Sometimes, the "tilt" of your profit objective is exactly parallel to one of your constraints. Imagine a flat ramp that aligns perfectly with a wall. In this case, any point along that wall is equally good. The solver will give you one answer, but it's important to know that multiple valid strategies might exist.

Visualizing the Difference Between Constrained and Unconstrained Optimization

Let's imagine a new problem. A company's profit is determined by two factors: ad_spend (x-axis) and product_quality_score (y-axis). The ideal, unconstrained, best profit is at the center of the contour plot presented in Figure 6-2. This is the true peak or the global maximum.

However, the company has two constraints:

➤ **A budget constraint:** They cannot spend more than $4,000 on ads. (A vertical line at x = 4,000.)

➤ **A resource constraint:** It's difficult to increase the quality score, so it can't go above 60. (A horizontal line at y = 60.)

These two lines create a constrained area, or a feasible region. They are only allowed to choose a solution that is to the left of the dashed line and below the solid line (see Listing 6-4).

LISTING 6-4: PROFIT MAXIMIZATION

```python
import numpy as np
import matplotlib.pyplot as plt

# --- Setup for the visualization ---
def calculate_profit(x, y):
    # A simple nonlinear function
    # The "true peak" (unconstrained optimum) is at (6000, 80)
    return -((x - 6000)**2 + (y - 80)**2) / 100000 + 50000

x = np.linspace(0, 10000, 200)
y = np.linspace(0, 120, 200)
X, Y = np.meshgrid(x, y)
Z = calculate_profit(X, Y)

# Unconstrained optimum
unconstrained_opt = (6000, 80)
# Constrained optimum
constrained_opt = (4000, 60)

# --- Plotting ---
plt.figure(figsize=(10, 8))
# Plot the profit contours
contours = plt.contour(X, Y, Z, levels=20, cmap='viridis')
plt.clabel(contours, inline=True, fontsize=8)

# --- Plot the Constraints ---
# 1. Budget Constraint (Vertical line at x=4000)
plt.axvline(x=4000, color='red', linestyle='--',
label='Budget Constraint (x <= 4000)')
# 2. Resource Constraint (Horizontal line at y=60)
plt.axhline(y=60, color='blue', linestyle='-',
label='Resource Constraint (y <= 60)')

# --- Plot the Optimums (Now with Diamonds 'D') ---
# Unconstrained (Red Diamond)
plt.plot(unconstrained_opt[0], unconstrained_opt[1], 'rD', marker-
size=10, label='Unconstrained Optimum (True Peak)')
# Constrained (Green Diamond)
plt.plot(constrained_opt[0], constrained_opt[1], 'go', marker-
size=10, label='Constrained Optimum (Best Possible)')

# --- Fill the Feasible Region ---
plt.fill_between(x, 0, 60, where=(x <= 4000), color='grey',
alpha=0.3, label='Feasible Region')

plt.title('Constrained vs. Unconstrained Optimization')
plt.xlabel('Ad Spend ($)')
plt.ylabel('Product Quality Score')
plt.legend()
plt.xlim(0, 10000)
plt.ylim(0, 120)
plt.grid(True)
plt.show()
```

The plot created by Listing 6-4 and shown in Figure 6-3 tells the whole story of the optimization. The *unconstrained optimum* (the diamond) is the true peak of the profit mountain. This is where the company wishes it could be, at an ad spend of $6,000 and a quality score of 80. The *constraints* (the lines) create a fence that represents the constraints. The company cannot go to the right of the dashed line or above the solid line. The *feasible region* (the shaded area) represents the feasible opportunity set for the company.

The *constrained optimum* (the dot) is the highest point inside the feasible region. The optimizer's job is to find this exact point. In this case, the best possible solution is to spend the entire $4,000 budget and hit the maximum possible quality score of 60. This is the core challenge of real-world business optimization. It's not about finding a theoretical "best," but about finding the best possible outcome given the real-world rules you have to play by. Recognizing and defining these constraints is the first and most critical step. This is the skill you will build on as you tackle more complex problems.

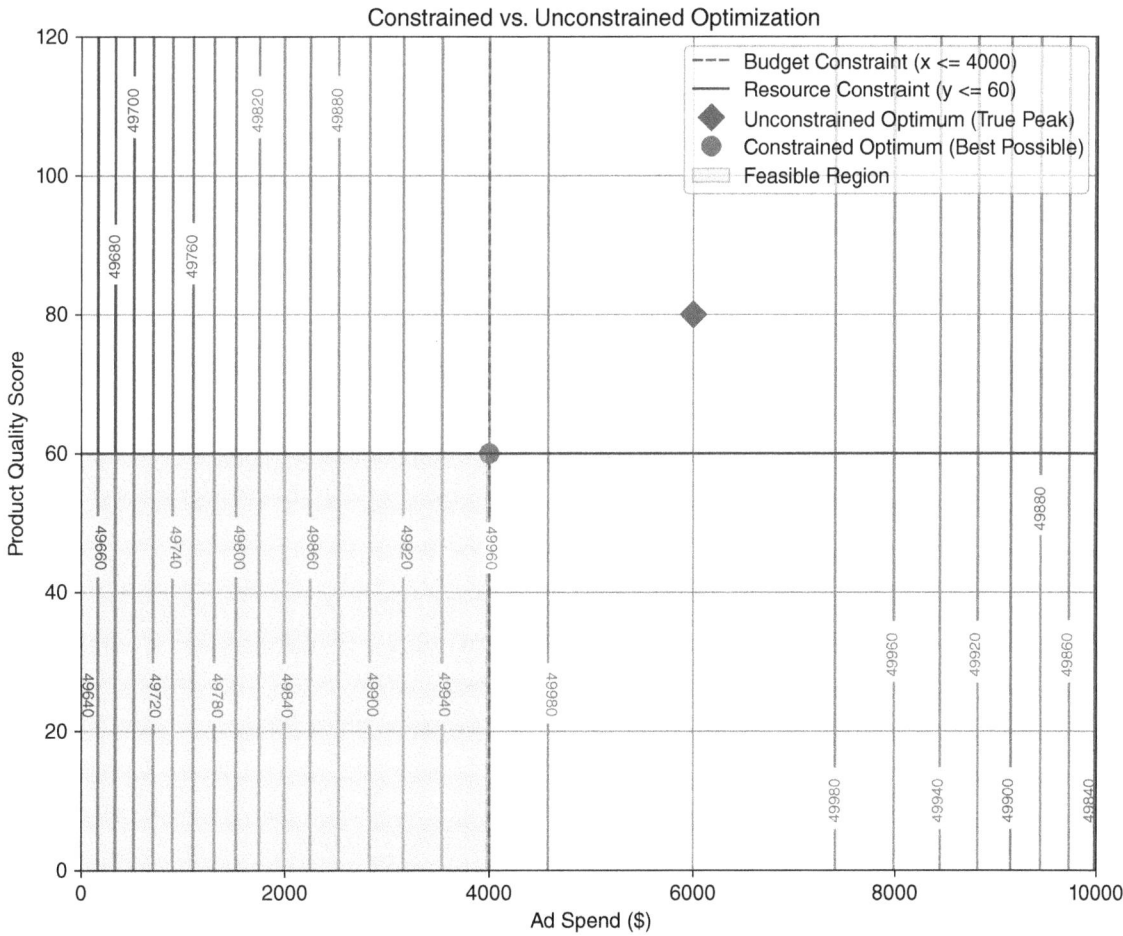

FIGURE 6-3: Constrained vs. unconstrained optimization plot.

REAL-WORLD APPLICATIONS

You are now ready to tackle the most realistic and powerful category of optimization problems. These often involve nonlinear relationships (curves, not straight lines) and strict constraints (rules you must follow). This section applies this optimization framework to three diverse real-world business challenges: minimizing risk in a stock portfolio subject to a required return, minimizing shipping costs in a complex supply chain network, and optimizing workforce scheduling when employees must work specific shifts. Each of these examples requires a slightly different approach and tool from the `scipy.optimize` toolkit, showcasing the versatility of the methods you've learned.

Portfolio Allocation

This section tackles a realistic and challenging type of constrained nonlinear optimization. Imagine you are a portfolio manager. You have four stocks to choose from, each with its own historical return and volatility. Your client has given you a very specific mandate: they want to achieve an expected annual return of at least 10%, while taking on the absolute minimum amount of risk. This is a perfect optimization problem for the framework. The first two steps are:

➤ **Objective:** Minimize risk. (Risk is measured by portfolio variance, a *nonlinear* function.)

➤ **Constraints:**

➤ The portfolio's expected return must be *greater than or equal to* 10%.

➤ All the money must be invested. (All weights or percentages must sum to 100%.)

➤ You can't short-sell (no negative weights).

To solve this, you need the most general-purpose optimizer, which is `scipy.optimize.minimize`. This is a powerful black box solver. You simply give it an objective function, an initial guess, and a set of constraints, and it finds the best solution.

Let's break down exactly how you should translate these requirements into the specific Python code that the minimize function expects. You need to start by setting up your data in Step 1.

Step 1: The Setup (Mock Data)

Assume you have four stocks (A, B, C, and D). You need two things: their expected returns and their covariance matrix. As a reminder, the covariance matrix measures how the stocks move together, as discussed in Chapter 4, "Linear Algebra for Business Finance."

Listing 6-5 creates these as the `exp_returns` and `cov_matrix` variables. Mock data is also assigned for four stocks across four years.

LISTING 6-5: SETTING THE ANNUAL RETURNS AND COVARIANCE MATRIX

```
import numpy as np
from scipy.optimize import minimize
```

```
# --- Asset Data (Mock Data) ---
# Expected annual returns for 4 assets
exp_returns = np.array([0.08, 0.12, 0.15, 0.07])

# Covariance matrix (how assets move together)
cov_matrix = np.array([
    [0.012, 0.002, 0.005, 0.001],
    [0.002, 0.025, 0.008, 0.003],
    [0.005, 0.008, 0.040, 0.006],
    [0.001, 0.003, 0.006, 0.010]
])

# Our target return
target_return = 0.10
```

Step 2: Define the Objective and Constraints

Now comes the crucial translation step. You need to express the investor's goal (minimize risk) and the rules they must follow (sum of weights = 1, return >= target) in a language that `scipy.optimize.minimize` understands.

Listing 6-6 first defines the objective function, `portfolio_variance`. This Python function takes a list of potential weights (the decision variables) and uses the covariance matrix to calculate the portfolio's total risk (the variance). The portfolio variance is the number the optimizer will try to make as small as possible. Next, you define a helper function, `portfolio_return()`, which calculates the expected return for a given set of weights.

You then need to define the constraints. Each constraint is represented as a dictionary. The first constraint, `con_sum_to_one`, ensures that all money is invested. The code uses `'type': 'eq'` for an equality constraint, telling the optimizer that the sum of the weights minus one must equal zero. The lambda function is just a compact way to define this rule.

The second constraint, `con_target_return`, ensures that you meet the minimum return goal. The code uses `'type': 'ineq'` for an inequality constraint, telling the optimizer that the calculated `portfolio_return` minus `target_return` must be greater than or equal to zero. Finally, the code bundles these constraint dictionaries into a list called `constraints` to pass to the solver.

LISTING 6-6: DEFINING THE OBJECTIVES AND CONSTRAINTS

```
# --- Step 2: Define Objective and Constraints ---

def portfolio_variance(weights):
    """Calculates portfolio variance (our objective to minimize)."""
    return weights @ cov_matrix @ weights

def portfolio_return(weights):
    """Calculates the portfolio's expected return."""
    return weights @ exp_returns

# Constraint 1: All weights must sum to 1
# We define it as sum(weights) - 1 = 0
con_sum_to_one = {'type': 'eq',
```

```
                        'fun': lambda weights: np.sum(weights) - 1}

# Constraint 2: Portfolio return must be >= target_return
# We define it as portfolio_return - target_return >= 0
con_target_return = {'type': 'ineq',
'fun': lambda weights: portfolio_return(weights) - target_return}

# Combine our constraints into a list
constraints = [con_sum_to_one, con_target_return]
```

Step 3: Define Bounds and Initial Guess

With the objective and constraints defined, the next step is to define two more pieces of setup. These are as follows:

➤ **Bounds:** You need to tell the optimizer the limits for each individual weight. Since you're not allowing short-selling (negative weights) and can't invest more than 100% in any single stock, each weight must be between 0 and 1. To accomplish this, the code in Listing 6-7 creates a tuple of (0, 1) pairs, one for each stock.

➤ **Initial guess (x0):** Remember the "local vs. global optima" problem? The minimize function needs a starting point for its search. A simple, unbiased starting point is an equal-weight portfolio, where you initially allocate the same percentage to each stock. The optimizer will then move from this guess toward the optimal solution.

Listing 6-7 includes the code for defining the bounds and applying the initial guess.

LISTING 6-7: DEFINING BOUNDS AND INITIAL GUESS

```
# --- Step 3: Define Bounds and Initial Guess ---

# Bounds: No short-selling (weights must be between 0 and 1)
bounds = tuple((0, 1) for _ in range(len(exp_returns)))

# Initial Guess: Start with an equal-weight portfolio
initial_guess = np.array([1/len(exp_returns)] * len(exp_returns))
```

Step 4: Run the Optimizer

The next step is where the magic happens! Listing 6-8 shows the call to the `scipy.optimize.minimize` function. You can see the function being fed all the pieces you've carefully prepared, which include:

➤ **fun=portfolio_variance:** The objective function you want to minimize.

➤ **x0=initial_guess:** The starting point for the search.

➤ **method='SLSQP':** You specify the algorithm to use *SLSQP* (Sequential Least Squares Programming) because it is a robust choice for handling both equality and inequality constraints in nonlinear problems.

➤ **bounds=bounds:** The (0, 1) limits for each weight.

➤ **constraints=constraints:** The list containing the rules (sum to 1, meet target return).

```
# --- Step 4: Run the Optimizer (Single Point) ---
result = minimize(
    fun=portfolio_variance,   # Function to minimize
    x0=initial_guess,         # Initial guess
    method='SLSQP',           # A good method for constrained problems
    bounds=bounds,
    constraints=constraints
)
```

The solver now takes over, algorithmically exploring different weight combinations within the allowed boundaries and constraints, and seeking the set of weights that produces the absolute lowest portfolio_variance.

Step 5: Interpret the Result

The optimizer returns its findings in a result object called `result`, as you can see in Listing 6-9. From this, you first check if `result.success` is true to ensure it found a valid solution. If it is true, you can assign a few variables based on the following:

> **result.x:** This array holds the optimal weights—the exact percentage to allocate to stock A, B, C, and D.

> **result.fun:** This holds the value of the objective function at the optimum—the minimum possible risk (variance) you can achieve.

You can then use the `portfolio_return()` function with the `optimal_weights` to double-check the portfolio's expected return, confirming that it meets the target.

```
# --- Step 5: Interpret the Single Result ---
if result.success:
    optimal_weights = result.x
    min_risk = result.fun
    calc_return = portfolio_return(optimal_weights)

    print("--- Optimal Portfolio Allocation (Single Target) ---")
    print(f"Target Return: {target_return*100:.1f}%")
    print("\nOptimal Weights:")
    for i, weight in enumerate(optimal_weights):
        print(f"  Stock {chr(65+i)}: {weight*100:.2f}%")

    print(f"\nCalculated Portfolio Return: {calc_return*100:.2f}%")
    print(f"Calculated Portfolio Risk (Variance): {min_risk:.4f}")

    # We will save this point to plot later
    single_point_risk = min_risk
    single_point_return = calc_return
else:
```

```
print("Optimization failed.")
print(result.message)
```

You can then print the results clearly. This gives the investor their precise, actionable plan: exactly how much to invest in each stock to hit their 10% return goal while minimizing risk. I also save the calculated risk and return for this specific portfolio so we can plot it later.

Based on the test data, the results are the following:

```
--- Optimal Portfolio Allocation (Single Target) ---
Target Return: 10.0%

Optimal Weights:
  Stock A: 30.96%
  Stock B: 26.78%
  Stock C: 16.89%
  Stock D: 25.37%

Calculated Portfolio Return: 10.00%
Calculated Portfolio Risk (Variance): 0.0074
```

Step 6: Generating the Efficient Frontier

Finding the optimal portfolio for one target return is useful, but the real power comes from seeing the trade-off across all possible returns. This is called the *efficient frontier.*

As shown in Listing 6-10, to generate the efficient frontier, you need to repeat the optimization process many times within a loop. You define a range of potential `target_returns` spanning from the lowest-returning stock, `exp_returns.min()`, to the highest, `exp_returns.max()`.

Inside the loop, for each target return, you do the following:

➤ Redefine the `con_target_return` constraint dictionary to use the current target return in the loop.

➤ Call `minimize` again with this updated constraint.

If the optimization is successful, you store the resulting minimum risk, `frontier_result.fun`, in the `frontier_risks` list.

After the loop finishes, `frontier_risks` and `frontier_returns` will contain pairs of risk/return values, each representing the single best portfolio for a given return level.

LISTING 6-10: GENERATING THE EFFICIENT FRONTIER

```
# --- Step 6: Generate the Efficient Frontier ---
# Now, we run the optimization many times for a range of targets

frontier_returns = np.linspace(exp_returns.min(), exp_returns.max(), 50)
frontier_risks = []

for target in frontier_returns:
    # Redefine the target return constraint for each loop
    con_target_return = {'type': 'ineq',
```

```
                          'fun': lambda weights, t=target: portfolio_
return(weights) - t}

    # Run the optimizer for the new target
    frontier_result = minimize(
        fun=portfolio_variance,
        x0=initial_guess,
        method='SLSQP',
        bounds=bounds,
        constraints=[con_sum_to_one, con_target_return]
    )

    if frontier_result.success:
        frontier_risks.append(frontier_result.fun) # Add the risk (variance)
    else:
        frontier_risks.append(np.nan) # Mark as failed
```

Step 7: Plotting the Efficient Frontier

With the efficient frontier generated, the next step is to visualize it. This is done using the code in Listing 6-11; the resulting plot is shown in Figure 6-4.

LISTING 6-11: VISUALIZING THE EFFICIENT FRONTIER

```
import matplotlib.pyplot as plt # --- Step 7: Visualize the Efficient Frontier ---
plt.figure(figsize=(10, 6))
plt.plot(frontier_risks, frontier_returns, 'g--', label='Efficient Frontier')
plt.plot(single_point_risk, single_point_return, 'ro', markersize=8,
label=f'10% Target Portfolio')

plt.title('Efficient Frontier for Portfolio Optimization')
plt.xlabel('Portfolio Risk (Variance)')
plt.ylabel('Expected Return')
plt.legend()
plt.grid(True)
plt.show()
```

The optimizer first did its job for the single target: it found the exact allocation of funds to achieve the 10% return goal with the lowest possible risk. The printout in Step 5 gives you a precise, actionable investment plan.

But the real power is in the visualization seen in Figure 6-4. The dashed line is the "efficient frontier," or a line representing the highest return and lowest risks for the different combinations of stocks. This is generated by running the optimization 50 times, each time asking for a different target return. Each point on that line represents the single best portfolio (the one with the lowest risk) for that level of return. Portfolios below the line are "inefficient." You could get the same return for less risk, or more return for the same risk. Portfolios to the right of the line are "impossible" given the set of assets. The circle is the specific 10% target portfolio you calculated first. Notice how it sits perfectly on the frontier, just as it should.

The resulting graph visually demonstrates the fundamental trade-off between risk and return. It shows the investor, for any desired level of return, the absolute minimum risk they must accept, allowing them to make a much more informed strategic decision about their investment goals.

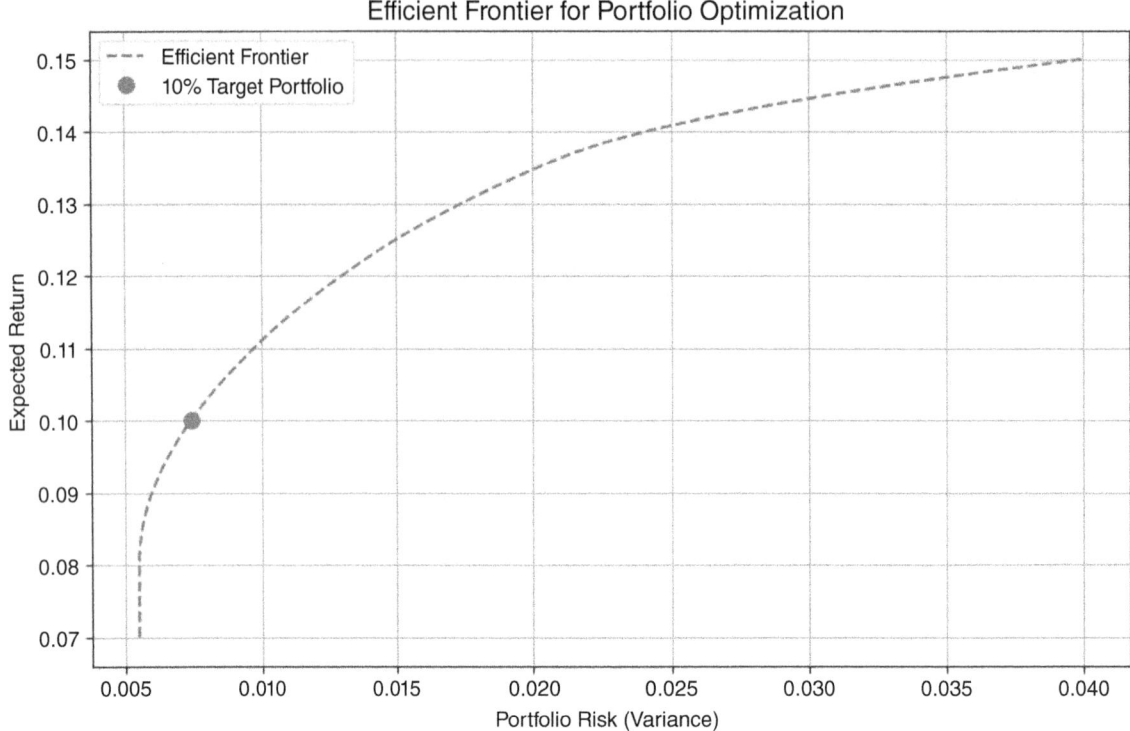

FIGURE 6-4: The efficient frontier plot for the portfolio optimization.

Supply Chain and Operations

This section returns to the world of linear programming (linprog) to solve another classic business problem. The Product Mix example was about allocating resources at a single location. Now, you see how to solve a problem between locations. More specifically, this is a transportation problem, a cornerstone of supply chain and logistics management, minimizing shipping costs.

Imagine you run a company with two warehouses and three retail stores. Each warehouse has a limited supply, and each store has a specific demand. Your goal is to find the cheapest way to get products from warehouses to stores.

Step 1: The Setup (Mock Data)

First, you need to lay out the data you have available.

- ➤ Supply:
  - ➤ Warehouse A has 1,000 units.
  - ➤ Warehouse B has 800 units.

- Demand:

  - Store 1 needs 500 units.

  - Store 2 needs 900 units.

  - Store 3 needs 400 units.

- (Total supply = 1,800, total demand = 1,800. They match.)

- Shipping costs (per unit):

  - From Warehouse A to Store 1: $2

  - From Warehouse A to Store 2: $4

  - From Warehouse A to Store 3: $5

  - From Warehouse B to Store 1: $3

  - From Warehouse B to Store 2: $6

  - From Warehouse B to Store 3: $3

The question is, how many units should you ship from each warehouse to each store to meet all demand, respect all supply limits, and do so at the minimum possible cost?

Step 2: Define Objective and Constraints

You now need to translate this business reality into the vectors and matrices that `linprog` understands. Listing 6-12 first defines the decision variables. There are six possible shipping routes (2 warehouses * 3 stores). You can flatten these into a single list of six variables: A->S1, A->S2, A->S3, B->S1, B->S2, and B->S3.

Next, you define the objective function. You want to minimize total cost, so you need to create a cost vector c containing the shipping prices for these six routes in order.

Finally, you define the constraints. In this problem, the constraints are equalities: the amount shipped from Warehouse A must exactly equal 1,000, and the amount received by Store 1 must exactly equal 500. Listing 6-12 sets this up using the A_eq matrix (defining which routes contribute to which constraint) and the b_eq vector (the actual supply/demand limits).

LISTING 6-12: SETTING UP THE TRANSPORTATION PROBLEM

```
import numpy as np
from scipy.optimize import linprog

# --- Step 1: Define the Problem ---

# Objective function (to be minimized):
# Cost = 2*x1 + 4*x2 + 5*x3 + 3*x4 + 6*x5 + 3*x6
c = [2, 4, 5, 3, 6, 3]

# Constraints (Equality): A_eq @ x = b_eq
```

```
# We have 5 constraints (2 supply, 3 demand)
# Variables: [x1, x2, x3, x4, x5, x6]
A_eq = [
    [1, 1, 1, 0, 0, 0],   # Wh. A Supply
    [0, 0, 0, 1, 1, 1],   # Wh. B Supply
    [1, 0, 0, 1, 0, 0],   # Store 1 Demand
    [0, 1, 0, 0, 1, 0],   # Store 2 Demand
    [0, 0, 1, 0, 0, 1]    # Store 3 Demand
]

b_eq = [
    1000, # Wh. A Supply
    800,  # Wh. B Supply
    500,  # Store 1 Demand
    900,  # Store 2 Demand
    400   # Store 3 Demand
]
# --- Define the Bounds ---
# We can't ship negative units
bounds = (0, None)
```

The final lines of Listing 6-12 set up logical limits on the variables. The only limit here is physical reality: you cannot ship a negative number of products.

Step 3: Optimizing

With the problem fully translated into c, A_eq, b_eq, and bounds, you can hand it off to the solver. The method='highs' is SciPy's recommended modern solver for these types of problems. The code in Listing 6-13 runs the solver and displays the results.

LISTING 6-13: RUNNING THE SOLVER AND DISPLAYING THE RESULTS

```
# --- Step 3: Run the Optimizer ---
result = linprog(c, A_eq=A_eq, b_eq=b_eq, bounds=bounds, method='highs')

# --- Step 4: Interpret the Results ---
if result.success:
    shipping_plan = result.x.reshape(2, 3) # Reshape 1x6 array into 2x3 matrix
    min_cost = result.fun

    print("--- Optimal Shipping Plan (Units) ---")
    print(f"          Store 1 | Store 2 | Store 3")
    print(f"Warehouse A: {shipping_plan[0, 0]:>7.0f} | {shipping_
plan[0, 1]:>7.0f} | {shipping_plan[0, 2]:>7.0f}")
    print(f"Warehouse B: {shipping_plan[1, 0]:>7.0f} | {shipping_
plan[1, 1]:>7.0f} | {shipping_plan[1, 2]:>7.0f}")
    print("\n---")
    print(f"Total Minimized Shipping Cost: ${min_cost:.2f}")

else:
    print("Optimization failed.")
    print(result.message)
```

The optimizer has produced a clear, actionable plan. To minimize costs, Warehouse A should handle all of Store 1's and Store 3's demand, plus a small part of Store 2's. Warehouse B should focus all its supply on fulfilling the rest of Store 2's large demand. You see this in the results:

```
--- Optimal Shipping Plan (Units) ---
            Store 1 | Store 2 | Store 3
Warehouse A:    100 |     900 |       0
Warehouse B:    400 |       0 |     400
---
Total Minimized Shipping Cost: $6200.00
```

Any other combination of shipments, while it might fulfill the demand, would result in a higher total cost. This is the kind of problem that saves large companies millions of dollars in logistics.

Integer Programming for Workforce Scheduling

So far, these optimization tools have served you well, but they share a common assumption: that the answers can be fractional. And for many problems, that's perfectly fine. It makes sense to allocate 38.38% of a portfolio or to understand that the theoretical profit peak is at 714.29 units. You can simply round to the nearest whole number.

However, sometimes you are confronted with decisions where rounding is not just inaccurate, it's not possible. Consider the following scenario.

Imagine you're a manager. Your optimization model tells you to dispatch 2.5 trucks to a new location. How do you dispatch half a truck? Or a model for a factory expansion that returns an answer of 0.6, when the only options are Yes (1) or No (0). The most common version of this problem is workforce scheduling. You simply cannot schedule 0.7 of an employee to cover a shift.

These situations require a new, more specialized tool: *integer programming* (IP). This is a branch of optimization where some, or all, of the decision variables are restricted to being whole numbers. Fortunately, the workhorse `scipy.optimize.linprog` has a parameter called *integrality* that allows you to solve exactly these kinds of puzzles.

Let's step into the shoes of a call center manager. They have a classic, real-world scheduling puzzle. They need to create a weekly staffing plan that meets the minimum number of employees required for each day, all while paying the lowest possible labor cost.

The core of the challenge lies in the shift structure. Each employee works for five consecutive days and then gets two days off. This means there are only seven possible "shift types" an employee can have (one starting on Monday, one on Tuesday, and so on). Each employee costs the company a flat $500 per week, regardless of which shift they are on.

The manager's daily demand, however, is not flat. The call volume fluctuates, requiring a different number of employees each day: 17 on Monday, 13 on Tuesday, 15 on Wednesday, 19 on Thursday, 17 on Friday, 10 on Saturday, and only 8 on Sunday.

The question is, how many employees should be hired for each of the seven shift types to meet this fluctuating daily demand at the absolute minimum cost?

Step 1: Formulating the Problem

This is an integer programming problem. You can't hire 2.5 people for the "Monday Start" shift. The answer must be a whole number.

First, you need to define the seven variables, which represent the seven decisions the manager has to make: x_mon (the number of employees starting on Monday), x_tue (starting on Tuesday), and so on, all the way to x_sun.

Second, you define the objective function. This is simple: you want to minimize the total cost. Since every employee costs $500, the function is Cost = 500*x_mon + 500*x_tue + ... + 500*x_sun.

Third, you build the constraints. This is the tricky part. You need to ensure that the number of people working on any given day is greater than or equal to that day's demand. Let's take Monday as an example. Who is on duty on a Monday?

➤ Employees who started their five-day shift on Monday (today).

➤ Employees who started on Sunday (day 2 of their shift).

➤ Employees who started on Saturday (day 3 of their shift).

➤ Employees who started on Friday (day 4 of their shift).

➤ Employees who started on Thursday (day 5 of their shift). Employees who started on Tuesday or Wednesday are on their days off.

So, the constraint for Monday becomes: x_mon + x_thu + x_fri + x_sat + x_sun >= 17

You then build a similar constraint for all seven days of the week, resulting in a system of seven inequalities. With this "blueprint" in hand, you can feed the problem to the solver. The resulting code is shown in Listing 6-14.

LISTING 6-14: INTEGER PROGRAMMING FOR WORK SCHEDULES

```
import numpy as np
from scipy.optimize import linprog

# --- Step 1: Define the Problem ---

# Objective function (Minimize Cost):
# Cost = 500*x1 + 500*x2 + ... + 500*x7
c = [500] * 7  # Cost is $500 for each of the 7 shift types

# Constraints (Left-hand side): A_ub @ x >= b_ub
# To make it "greater than", we multiply A and b by -1
# A_ub @ x <= b_ub  --->  -A_ub @ x >= -b_ub
#
#       Mon Tue Wed Thu Fri Sat Sun (Shifts Starting)
# Mon:   1   0   0   1   1   1   1    >= 17
# Tue:   1   1   0   0   1   1   1    >= 13
# Wed:   1   1   1   0   0   1   1    >= 15
# Thu:   1   1   1   1   0   0   1    >= 19
```

```
# Fri:   1   1   1   1   1   0   0   >= 17
# Sat:   0   1   1   1   1   1   0   >= 10
# Sun:   0   0   1   1   1   1   1   >= 8

A_ub = -np.array([
    [1, 0, 0, 1, 1, 1, 1],   # Mon
    [1, 1, 0, 0, 1, 1, 1],   # Tue
    [1, 1, 1, 0, 0, 1, 1],   # Wed
    [1, 1, 1, 1, 0, 0, 1],   # Thu
    [1, 1, 1, 1, 1, 0, 0],   # Fri
    [0, 1, 1, 1, 1, 1, 0],   # Sat
    [0, 0, 1, 1, 1, 1, 1]    # Sun
])

b_ub = -np.array([17, 13, 15, 19, 17, 10, 8]) # Daily minimums

# --- Step 2: Define Bounds and Integrality ---
bounds = (0, None) # Can't hire negative people
# Tell the solver all 7 variables must be integers
integrality = [1] * 7

# --- Step 3: Run the Optimizer ---
result = linprog(c, A_ub=A_ub, b_ub=b_ub, bounds=bounds, integrality=integrality,
method='highs')

# --- Step 4: Interpret the Results ---
if result.success:
    schedule = result.x
    min_cost = result.fun
    total_employees = np.sum(schedule)
    days = ['Monday', 'Tuesday', 'Wednesday', 'Thursday', 'Friday', 'Saturday',
'Sunday']

    print("--- Optimal Staffing Schedule ---")
    print("Employees starting on:")
    for i, num in enumerate(schedule):
        print(f"  {days[i]:<10}: {num:.0f}")

    print("\n---")
    print(f"Total Employees Hired: {total_employees:.0f}")
    print(f"Total Minimized Weekly Cost: ${min_cost:.2f}")
else:
    print("Optimization failed.")
    print(result.message)
```

Step 2: Interpreting the Results

The results of the solver are as follows:

```
--- Optimal Staffing Schedule ---
Employees starting on:
Monday : 6
Tuesday : 4
Wednesday : 0
Thursday : 7
```

```
Friday : 0
Saturday : 0
Sunday : 2
--- Total Employees Hired: 19
Total Minimized Weekly Cost: $9500.00
```

This is a powerful, non-obvious solution. The solver determined that the most cost-effective way to meet the fluctuating daily demand is to hire a total of 19 employees, starting them on Monday, Tuesday, Thursday, and Sunday. No employees should start their shifts on Wednesday, Friday, or Saturday.

This plan meets all the minimum staffing requirements for each day at the absolute lowest possible cost. This type of integer programming is a fundamental tool for managers in operations, HR, and logistics.

SUMMARY

This chapter has been a journey, one that began with a single question from the last chapter: "how do you find the exact peak of the profit curve?" You've traveled from that one simple query to a complete framework for solving complex business problems. You've seen that every optimization problem, no matter how intimidating, can be broken down into two core components: an objective (what you want) and its constraints (the rules you must follow).

Using the `scipy.optimize` toolkit, you began by answering that initial question, finding the simple, unconstrained peak of the profit curve. From there, you moved into the world of real-world limits, using linear programming to solve a "product mix" problem where you discovered a surprising, non-obvious solution that maximized profit by focusing on a lower-margin product. You then explored the fundamental difference between unconstrained and constrained problems, visualizing the "feasible region" that defines the boundaries of all real-world decisions.

With this foundation, you were ready to tackle truly complex nonlinear challenges, building an optimal financial portfolio and generating the efficient frontier to visualize the trade-off between risk and return. Your toolkit expanded again to solve tangible operations problems, including a complex supply chain logistics puzzle and an intricate integer programming problem for scheduling staff, where fractional answers simply weren't an option. Finally, you brought all these concepts together by building a sophisticated pricing model that optimized for two interacting products.

You now have a proven framework for translating almost any business challenge, from finance to operations to marketing, into a solvable model. You have officially moved from analysis to prescription, and you are equipped with the tools to find not just a good solution, but the "best" one.

CONTINUE YOUR LEARNING

For readers who want to dive deeper into the powerful optimization tools discussed in this chapter, the following official resources are highly recommended:

➤ **SciPy Optimization User Guide:** The definitive guide to all solvers available in the library, including many advanced algorithms not covered here. `https://docs.scipy.org/doc/scipy/tutorial/optimize.html`

➤ **Linear programming with SciPy:** Detailed documentation on the linprog function, including advanced options for the "highs" solvers used in the examples. `https://docs.scipy.org/doc/scipy/reference/generated/scipy.optimize.linprog.html`

➤ **General minimization:** In-depth details on the minimize function used for the portfolio examples, including the different algorithms available (like SLSQP and trust-constr). `https://docs.scipy.org/doc/scipy/reference/generated/scipy.optimize.minimize.html`

➤ **CVXPY:** If you are interested in a more advanced, algebraic way to model complex convex optimization problems, CVXPY is the industry standard in Python. `https://www.cvxpy.org/`

Probability and Statistics for Business Analytics

In the previous chapter, the optimization models led to a powerful, precise answer: the optimal production quantity was 714.29 units, yielding a maximum profit of $10,714.29. But this answer was built on a critical assumption: that the inputs (cost, demand, etc.) were fixed, known numbers.

Business reality is much more complex. In that reality, the demand isn't exactly 714, it's around 714, and it could be 650 on a slow day or 800 during a sales rush. The variable costs aren't exactly $50, they fluctuate. This chapter is about introducing the framework for making smart decisions in the face of this real-world uncertainty. It moves from the what-if analysis of the previous chapters to a what's-likely analysis.

To do this, you will use a familiar and new set of Python libraries. You'll use `numpy.random` to simulate this randomness, `scipy.stats` to run formal statistical tests, and `statsmodels` to build powerful models that explain why the numbers change.

THE PYTHON STATISTICS ECOSYSTEMS

Before diving into solving problems, it's helpful to know what tools are available. Unlike optimization, which is dominated by `scipy.optimize`, the statistics landscape is a collaboration between several key libraries, each with a specific job:

➤ **NumPy (`numpy.random`):** NumPy's core library provides the array structures, but its random submodule is the tool for creating data. You'll see how to use it to simulate sales, model customer behavior, and generate the random inputs for the risk analysis.

➤ **SciPy (`scipy.stats`):** This is the primary statistical toolkit. Once you have data (either real or simulated), you use `scipy.stats` to analyze it. It's packed with hundreds of functions for describing distributions, calculating confidence intervals, and, most importantly, running hypothesis tests like the t-test.

➤ **Statsmodels (`statsmodels.api`):** To move beyond simple tests and find the relationship between variables (like how ad spend affects sales), you'll use `statsmodels`. It's famous for its ability to perform sophisticated linear regressions and produce detailed summary tables that help elucidate what's driving your business.

➤ **Pandas:** This is the container that holds everything together. While not a statistics library itself, Pandas DataFrames are the standard way to load, clean, and organize the real-world data that you will feed into SciPy and Statsmodels.

> **NOTE** These libraries are free and can be added using pip or conda.

The workflow in this chapter will generally follow this path: you will use `numpy.random` to simulate realistic data, `scipy.stats` to test it, and `statsmodels` to build predictive models from it.

While these packages serve as the foundation for probability and statistics, the landscape of statistical packages can be very broad depending on the problem you are trying to solve. We cannot cover all of the possible options in this chapter alone, so the "Continue Your Learning" section at the end of this chapter has more information on additional statistical libraries.

UNDERSTANDING RANDOM VARIABLES AND DISTRIBUTIONS IN BUSINESS CONTEXTS

Before you can model uncertainty, you need a language to describe it. In statistics, we do this with random variables (a variable that can take on a range of values, not just one) and distributions (the shape of that randomness).

Imagine a random variable called daily sales. You know that tomorrow's daily sales probably won't be $10, and it probably won't be $10 million, but it could be any number within a realistic range.

On Monday, daily sales are $480. On Tuesday, daily sales stop at $510, and it continues to $495 on Wednesday. Each day, it takes on a new, slightly different value based on all the random factors of the real world. Those factors that make daily sales move up and down, that's what makes daily sales a random variable.

If you just looked at one or two days, you wouldn't see a pattern. If you recorded and plotted this value for 1,000 days and grouped them together, you'd see a shape emerge such as the one shown in Figure 7-1. This shape is the distribution. You'd likely see a big pile of values clustered around the average and fewer and fewer values as you get farther out. This shape is the visual record of all the variable's past movements. It also tells you the probability of where the variable might land next.

The specific shape shown in Figure 7-1 is called the *normal distribution*, sometimes also referred to as a *bell curve*. It represents an empirical distribution because it uses real data to create the distribution. Each bar represents the proportion of observations that are around that value, and the dashed line shows the general shape carted by the bars.

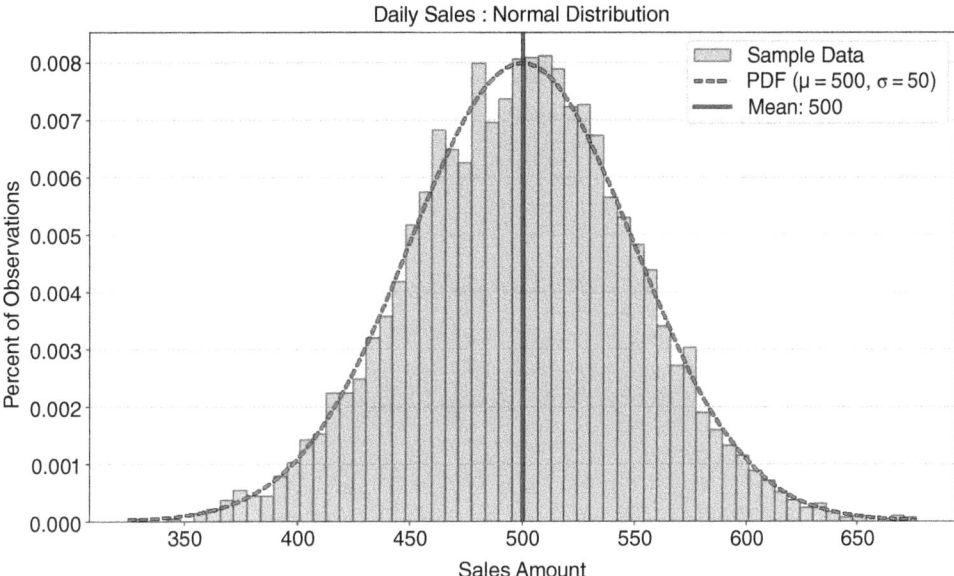

FIGURE 7-1: Distribution of daily sales.

You can see that this fits with what you generally know to be true in real life. Usually there is a baseline level of sales in business—sometimes that value goes up due to the holidays, and other times it goes down due to other events. The plot aligns with that reality, you can see that the daily sales observations around 500 shows up most frequently, and it is rare to see 400- or 600-dollar sales days.

Now that you have a general idea of what distributions and random variables look like, the next section explores how different data can create different distributions.

Discrete vs. Continuous Distributions

In business, we generally work with two key types of variables:

➤ **Discrete variables:** The variable can only take on specific, countable values. For example, a customer either converted (1) or did not convert (0). Another example is the number of items sold, which can be 1, 2, 3, and so on (you can't sell 1.5 t-shirts, for example).

➤ **Continuous variables:** The variable can take on any value within a range. Examples include the exact time a user spends on your site (e.g., 12.534 seconds) or the exact amount of a sale.

This distinction between discrete and continuous variables is critical because it dictates how you model them. A discrete variable, like number of items sold, moves in distinct, whole-number steps (1, 2, 3). A continuous variable, like time spent on site, is measured, not counted, and can take on any fractional value (e.g., 12.51 or 12.52 seconds).

Once you've identified your variable's type, the next logical step is to understand its behavior. Not all data fits the shape of the normal distribution. If you let a random variable (such as the number of items purchased) take on thousands of values, you would see a new distinct shape emerge.

The Most Common Business Distributions

Now that you can classify your variables, you can explore the most common distributions used to model them. While there are dozens of statistical distributions, most business analysis will be driven by four key types. The key types discussed here are the normal distribution, binomial distribution, the uniform distribution, and the Poisson distribution. This section investigates the different distributions using NumPy and Matplotlib to plot their classic distributions.

The Normal Distribution (the "Bell Curve")

The normal distribution is the most important continuous distribution in statistics. It's the default model for countless real-world phenomena that cluster around an average, from product sales to variations in manufacturing. It's defined by its mean (the average, or center) and its standard deviation (spread).

Listing 7-1 generates and visualizes a standard normal distribution (with a mean of 0 and a standard deviation of 1) and shows a classic shape.

LISTING 7-1: VISUALIZING THE NORMAL DISTRIBUTION

```
import numpy as np
import matplotlib.pyplot as plt

# 1. Normal Distribution (Continuous)
mu, sigma = 0, 1 # mean and standard deviation
s = np.random.normal(mu, sigma, 10000)

# --- Visualization ---
plt.figure(figsize=(8, 5))
plt.hist(s, bins=30, density=True, edgecolor='black', alpha=0.7)
plt.title('Normal Distribution (Bell Curve)')
plt.xlabel('Value')
plt.ylabel('Probability Density')
plt.grid(True)
plt.show()
```

Listing 7.1 uses `np.random.normal(mu, sigma, 10000)` to generate 10,000 random numbers drawn from a normal distribution with a mean of 0 and a standard deviation of 1. It then uses the `bins=30` argument to divide the range of data values into 30 equal-width intervals (or buckets). This groups the continuous data points into discrete chunks to visualize the frequency distribution. It then uses `plt.hist()` to plot these numbers. The `density=True` argument normalizes the histogram, so the area under the curve seen in Figure 7-2 sums to 1, representing 100% of outcomes.

The resulting histogram in Figure 7-2 clearly shows the bell curve. Most values are clustered around the mean (0), and the probability of seeing a value tapers off as you get farther away. This is the shape of many business metrics, as you'll see in the next example.

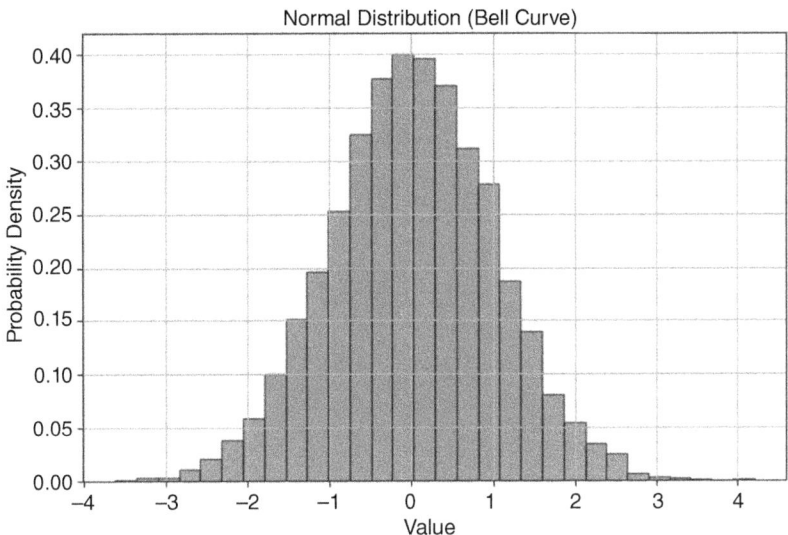

FIGURE 7-2: The normal distribution.

The Binomial Distribution

The *binomial distribution* is the most important discrete distribution for modeling a sequence of trials. It helps you model the number of successes you'll get in a set number of trials. It is defined by n (the number of trials) and p (the probability of success for each trial).

Let's visualize a common scenario: if you flip a fair coin 100 times (n = 100, p = 0.5), what is the likely number of heads? Listing 7-2 simulates this 10,000 times to build a distribution.

LISTING 7-2: VISUALIZING THE BINOMIAL DISTRIBUTION

```
import numpy as np
import matplotlib.pyplot as plt

# 2. Binomial Distribution (Discrete)
n, p = 100, 0.5  # number of trials, probability of success
s = np.random.binomial(n, p, 10000)

# --- Visualization ---
plt.figure(figsize=(8, 5))
plt.hist(s, bins=20, density=True, edgecolor='black', alpha=0.7)
plt.title('Binomial Distribution (100 Coin Flips)')
plt.xlabel('Number of Heads (Successes)')
plt.ylabel('Probability')
plt.grid(True)
plt.show()
```

Listing 7-2 uses np.random.binomial(n, p, 10000). This function performs the 100 coin flips experiment 10,000 times and records the number of successes for each experiment. The resulting histogram plots the outcomes in Figure 7-3.

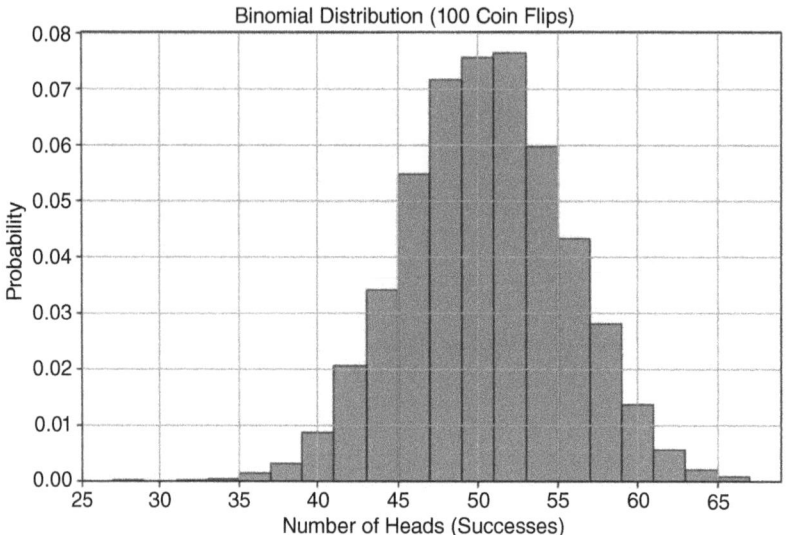

FIGURE 7-3: The binomial distribution.

Outside of coins, where does this distribution show up in the business context? Imagine that you sent a marketing email to 1,000 people (n = 1,000). You have a 3% click-through rate (p = 0.03). What's the likely range of people who will click? This model would tell you how many clicks to expect (around 30) and the probability of a great day (50 clicks) and a terrible day (10 clicks). This is the statistical engine behind all A/B testing and conversion rate analysis.

The Uniform Distribution

The *uniform distribution* is the simplest of all. It models a continuous situation where all outcomes in a given range are equally likely. It is defined by a low and high value.

Listing 7-3 illustrates a uniform distribution by using np.random.uniform(-1, 1, 10000) to generate 10,000 random numbers, where any value between –1 and 1 has an equal chance of being chosen.

LISTING 7-3: VISUALIZING THE UNIFORM DISTRIBUTION

```
import numpy as np
import matplotlib.pyplot as plt

# 3. Uniform Distribution (Continuous)
s = np.random.uniform(-1, 1, 10000)

# --- Visualization ---
plt.figure(figsize=(8, 5))
plt.hist(s, bins=30, density=True, edgecolor='black', alpha=0.7)
plt.title('Uniform Distribution (Equal Chance)')
plt.xlabel('Value')
```

```
plt.ylabel('Probability Density')
plt.grid(True)
plt.show()
```

As you can see in Figure 7-4, the histogram is flat. This is the shape of pure, unbiased randomness within a defined range. You can see this type of distribution in business as well. Imagine that your supplier says the delivery will arrive in 5–10 days. You have no other information. When modeling this in a simulation, you would use a uniform distribution (low = 5, high = 10). It's the most unbiased way to model uncertainty when you only know the minimum and maximum possible values.

The Poisson Distribution

The *Poisson distribution* is another essential discrete distribution. Instead of modeling successes in n trials like the binomial, Poisson models the number of events that occur in a fixed interval of time or space. It is defined by a single parameter, *lam* (lambda), which is the average number of events in that interval.

This is the perfect model for your number of products sold. Let's say a small e-commerce site averages 10 sales per hour. You want to simulate the number of sales they might get in any given hour. You can set up this model in Listing 7-4.

LISTING 7-4: VISUALIZING THE POISSON DISTRIBUTION

```
import numpy as np
import matplotlib.pyplot as plt

# 4. Poisson Distribution (Discrete)
avg_events_per_hour = 10
s = np.random.poisson(avg_events_per_hour, 10000)

# --- Visualization ---
plt.figure(figsize=(8, 5))
# We can use np.bincount to get the frequency of each integer
counts = np.bincount(s)
plt.bar(range(len(counts)), counts, align='center', edgecolor='black', alpha=0.7)
plt.title(f'Poisson Distribution (Avg = {avg_events_per_hour} events/hour)')
plt.xlabel('Number of Sales in One Hour')
plt.ylabel('Frequency (out of 10,000 simulations)')
plt.grid(axis='y')
plt.xlim([0, 25]) # Truncate x-axis for readability
plt.show()
```

This example uses `np.random.poisson(avg_events_per_hour, 10000)`. This simulates 10,000 different hours and records the number of sales (events) that occurred in each. Instead of `plt.hist`, it uses `plt.bar` with `np.bincount` to create a clean bar chart, which is more appropriate for integer data like this.

As you can see in Figure 7-5, the most likely outcome is the average of 10 sales. However, the distribution is not symmetrical (it's "skewed right"). This model shows that it's common to have 8 or 12 sales, but very rare to have 20, and impossible to have negative sales.

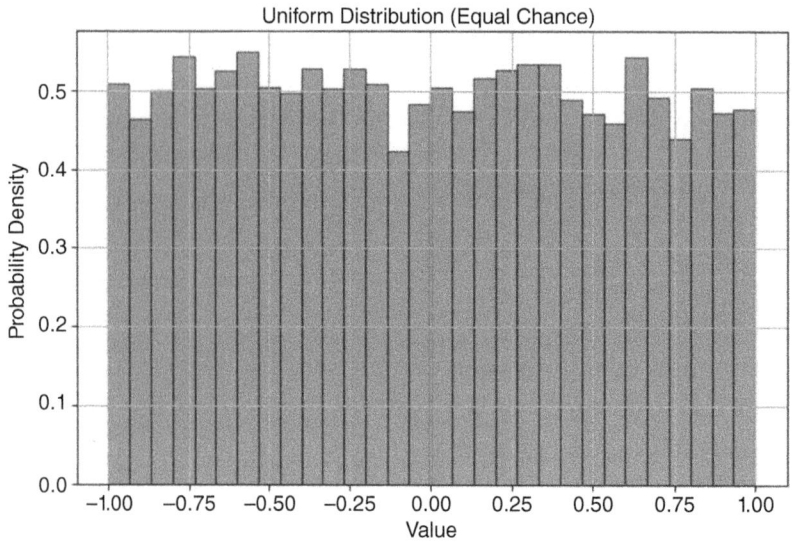

FIGURE 7-4: The uniform distribution.

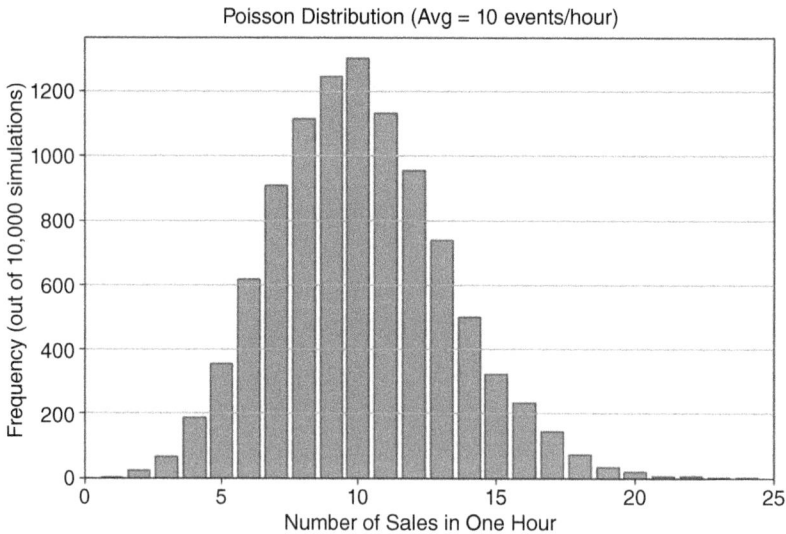

FIGURE 7-5: A histogram of simulated hourly sales.

HYPOTHESIS TESTING

Now that you can describe data with distributions, how do you make a decision with it or determine if a change makes a difference? This is the domain of *hypothesis testing,* the rigorous framework for answering the question: Did my change have a real effect, or did I just get lucky?

Imagine a classic business scenario: You run an A/B test on your website.

➤ **Version A (Control):** The current design.

➤ **Version B (Test):** A new design with a bigger Buy Now button.

You run the test for a week. The results come in: Version A had a 10% conversion rate, and Version B had a 12% conversion rate. It *looks* like Version B is the winner. But is it? Or is that 2% difference just random noise, the same way flipping a coin 10 times might give you six heads one day and four the next? You need a way to separate the signal from the noise. That is where hypothesis testing comes in.

Hypothesis testing works by setting up a debate between two competing viewpoints:

➤ The **null hypothesis** assumes that nothing interesting happened. In this example, the null hypothesis is: There is no real difference between Version A and Version B. Any difference you see is just random luck. In science and business, we always assume the null hypothesis is right until proven otherwise.

➤ The **alternative hypothesis** is what we are trying to prove. In this example, it is: There is a real difference; Version B is statistically better than Version A.

To settle the debate of which hypothesis is correct, you use a test statistic. This calculation effectively measures how far your observed data is from what the null hypothesis predicted. The two most common types are as follows:

➤ **z-value (z-statistic):** This measures how many standard deviations your result is from the mean. It is typically used when you have a large sample size (usually over 30) and you know the population's variance.

➤ **t-value (t-statistic):** This is similar to the z-value but is adjusted for smaller sample sizes or when the population variance is unknown. It is slightly more conservative, accounting for the higher uncertainty in small datasets.

The decision to reject or fail to reject the null hypothesis using t-statistics (or z-statistics) and p-values is a two-step process that is logically equivalent but approaches the problem from different angles.

Test Statistics

Think of the test statistic as a standardized measure of how far your sample result is from what you expected under the null hypothesis. We use test statistics to make decisions. Another way you can think about them is we are projecting our data onto a typical distribution. When we make a decision, we are just looking at how abnormal our result is relative to the typical distribution. You can determine if your null hypothesis is incorrect by following a simple three-step process.

Step 1: Set a Critical Value

Before running the test, you decide on a critical region based on your confidence level (usually 95%). For a standard two-tailed test, this critical value is often around 90% (for z-scores).

Step 2: Calculate the Stat

You calculate a t-score (for smaller samples) or z-score (for large samples). This number represents how many standard deviations your result is away from the null hypothesis mean.

Step 3: Make a Decision

Using your calculations, make a decision based on the following:

➤ If your calculated statistic is more extreme than the critical value (e.g., 2.5 > 1.96), your result falls into the "rejection region." You reject the null hypothesis.

➤ If it is less extreme, you fail to reject the null hypothesis.

Notice that we never say we accept the null hypothesis. This is a subtle but critical distinction in statistics. Failing to reject the null hypothesis doesn't mean you've proven it's true; it simply means that you haven't found enough evidence to prove it's false. Think of it like a court trial: a defendant is found not guilty, never innocent. You assume the null is true until the evidence (the data) forces you to abandon it. If the p-value is high, the evidence was just too weak to convict.

The p-value

While test statistics are a powerful tool, calculating them is only half the battle. To make a final decision, you traditionally have to look up a "critical value" in a statistical table, which can feel like an extra, cumbersome step. This is where the p-value comes in. It simplifies the entire process by converting your test statistic into a single, intuitive probability.

Think of the p-value as a direct translation of your t-score or z-score into a language of likelihood. Instead of asking, "Is my score of 2.5 greater than the critical value of 1.96?" the p-value answers a more direct question: If the skeptic were right and there was truly no effect, what are the odds I would see data this extreme just by random luck?

This connection is mathematically precise. A high-test statistic (representing a large difference from the norm) will always result in a low p-value (representing a low probability of luck). This means you can skip the lookup tables entirely. You simply compare your p-value directly to your risk tolerance. For most research studies, for example pharmaceutical trials, the comparison p-value is 0.05. A simple rule of thumb is as follows:

➤ **If $p < 0.05$:** The probability of this being luck is so low that you reject the skeptic's view. (This is the same as your test statistic falling into the rejection region.)

➤ **If $p > 0.05$:** The probability of luck is reasonably high, so you cannot rule it out. You fail to reject the null hypothesis.

By focusing on the p-value, you get the same rigorous conclusion as the t-stat or z-stat method, but with a number that is much easier to interpret and explain to stakeholders.

The A/B Test

This section simulates a problem where you can use hypothesis testing. For this, you will simulate an A/B test. You'll create two datasets representing conversion rates for Website A and Website B. The data will be slightly skewed toward Website B, and then you'll use `scipy.stats` to see if a t-test can detect this difference. Even though you are not working with real data and you know the data ahead of time, you can see how this might be the type of problem businesses could face on any given day.

Listing 7-5 uses `np.random.binomial` to simulate the user behavior by creating two datasets of 1,000 website visitors each (n_samples). For Website A, we simulate conversions using a binomial distribution with a 10% success rate, while for Website B, we use a 12% rate. This step allows you to generate realistic, messy data where you know the "ground truth" (that B is better) so you can test if your statistical tools can successfully detect it.

LISTING 7-5: RUNNING A T-TEST ON A/B DATA

```python
import numpy as np
from scipy import stats

# 1. Simulate our data
# We use the Binomial distribution (1=converted, 0=not)
# We simulate 1,000 visitors for each version
n_samples = 1000
# Website A has a true rate of 10%
conversions_a = np.random.binomial(1, 0.10, n_samples)
# Website B has a true rate of 15%
conversions_b = np.random.binomial(1, 0.15, n_samples)

# 2. Calculate the observed conversion rates
rate_a = np.mean(conversions_a)
rate_b = np.mean(conversions_b)

print(f"Observed Conversion Rate A: {rate_a:.1%}")
print(f"Observed Conversion Rate B: {rate_b:.1%}")

# 3. Run the Hypothesis Test (t-test)
# We use 'ttest_ind' because we are comparing two INDEPENDENT groups.
# We set equal_var=False (Welch's t-test) because in the real world,
# we can rarely assume two groups have the exact same variance.
t_stat, p_value = stats.ttest_ind(conversions_a, conversions_b, equal_var=False)

print(f"\nP-Value: {p_value:.4f}")

# 4. Interpret the result
if p_value < 0.05:
    print("Result: Statistically Significant! (Reject the Null Hypothesis)")
    print("We are confident the difference is real.")
else:
    print("Result: Not Significant. (Cannot Reject the Null Hypothesis)")
    print("The difference could just be random noise.")
```

After setting up a simulation of data, we calculate the observed conversion rates by taking the mean of each dataset. This mimics the real-world dashboard view a manager would see, showing the actual percentage of simulated visitors who converted in this specific experiment, which will likely hover around, but not exactly match, the 10% and 15% settings due to randomness.

Running the code in Listing 7-5 will produce the following results:

```
Observed Conversion Rate A: 10.3%
Observed Conversion Rate B: 15.0%

P-Value: 0.0016
Result: Statistically Significant! (Reject the Null Hypothesis)
We are confident the difference is real.
```

The core of the analysis happens when we run the hypothesis test. We use `stats.ttest_ind` to compare the two independent groups of visitors. Crucially, we set `equal_var=False` to perform Welch's t-test, a robust method that does not assume both groups have the same variance, which is a safer and more professional assumption for real business data. This function calculates the t-statistic and, most importantly, the p-value. Finally, we interpret this p-value using a standard 5% significance threshold (0.05). If the p-value is below 0.05, the code prints that the result is statistically significant, meaning the probability of seeing this difference by sheer luck is so low (less than 5%) that we reject the null hypothesis and conclude the lift is real. If it's above 0.05, we report that the result is not significant, meaning we cannot rule out the possibility that the difference is just random noise.

From the results, you can see that the p-value is 0.0016, meaning that Website A and Website B are not the same. Looking at the means again, you can see that Website B clearly has a superior conversion rate.

Confidence Intervals: The Other Side of the Coin

While the p-value provides a binary answer, Yes or No, business leaders usually require more nuance. A simple, Version B is better, is often insufficient for making high-stakes decisions. Stakeholders typically ask a different, more quantitative question: How much better is it?

To answer this, you need to move beyond simple hypothesis testing to calculating a *confidence interval*. Instead of a single point estimate, a confidence interval provides a range of values, giving context to your findings by quantifying the uncertainty. Rather than just reporting that Version B has a 12% conversion rate, you can say that you are 95% confident that the true conversion rate for Version B lies between 10.5% and 13.5%. This range is far more valuable for risk assessment and ROI calculation.

You can calculate this interval using `scipy.stats`. Listing 7-6 defines the desired confidence level (typically 95%) and uses the standard error of the mean (SEM) to construct the interval around the sample mean. The `stats.t.interval` function handles the heavy lifting, using the t-distribution to determine the appropriate width of the interval based on the sample size and variability.

LISTING 7-6: CALCULATING A CONFIDENCE INTERVAL

```
confidence_level = 0.95
degrees_freedom = n_samples - 1
sample_mean = np.mean(conversions_b)
# Calculate the Standard Error of the Mean (SEM)
sample_standard_error = stats.sem(conversions_b)

# Create the interval based on the t-distribution
confidence_interval = stats.t.interval(confidence_level, degrees_freedom, sample_
mean, sample_standard_error)

print(f"95% Confidence Interval for Website B: {confidence_interval[0]:.1%} to
{confidence_interval[1]:.1%}")
```

The result of running the code in Listing 7-6 is as follows:

```
95% Confidence Interval for Website B: 12.8% to 17.2%
```

This output gives managers a realistic range of expectations. If the interval was wide, say 8–16%, it would indicate a high degree of uncertainty, suggesting that you might need more data before making a decision. Conversely, a narrow interval like 11.8–12.2% indicates high precision. This context is essential for calculating the ROI of any business decision, allowing leaders to prepare for both the best-case and worst-case scenarios.

LINEAR REGRESSION

So far, you've seen how to used statistics to describe uncertainty with distributions and to make yes/no decisions with hypothesis tests. You've answered questions like: Is Version B better than Version A? But you haven't answered a much more powerful question: Why?

In a real business, sales don't just change randomly. They are driven by forces we can control: ad spend, pricing, promotions, and more. The final step in your core statistics toolkit is moving from simple comparison to explanation and prediction.

This is the job of *regression analysis*. Instead of just asking if a change had an effect, you build a model to quantify that effect. You want to answer the most common and valuable questions in business:

> How much does ad spend really affect sales?

> If you spend $1 more on marketing, how many dollars in sales will you get back?

> How sensitive are your customers to price? If you raise the price by $1, how many sales will you lose?

To do this, you'll use a powerful tool called *multivariate linear regression*. Think of the simple trend line we created in the last section:

$$Sales = B_0 + B_1 (Time)$$

That was a good start, but it's a simple model. Time isn't a business strategy. It doesn't cause sales to go up; it's just a placeholder for the real, underlying drivers of growth. Now, we are going to build a much smarter model by replacing the generic *Time* variable with the actual levers managers can pull: ad spend and price. You can build a model that looks like this:

$$Sales = B_0 + B_1 * (AdSpend) + B_2 * (Price)$$

This equation is the core of the analysis. B_0 is the intercept, representing baseline sales if you spent $0 on ads and your price was $0. The B_1 and B_2 terms are the coefficients, and they are the treasure you are looking for. B_1 will tell you the exact dollar-for-dollar power of your advertising, and B_2 will tell you the precise impact of your price on customer demand.

The computer finds these perfect coefficients using a method called *ordinary least squares* (OLS). Imagine a scatterplot of all your data, a cloud of points in space, where each point's position is defined by its ad spend, price, and sales. OLS is a mathematical process that finds the single best-fit plane (think of it as a flat sheet of paper) that slices through that cloud of data. This visualization would look something like Figure 7-6.

Best fit means it finds the plane that has the smallest possible total error. The error (also called a *residual*) is the vertical distance between each individual data point and the model's plane. In Figure 7-6, the error is the distance between the point and the solid line, highlighted by a dotted line. OLS calculates this distance for all 100 points, squares them, and then adds them all up. Then, using the optimization methods from the previous chapter, it finds the one unique plane that makes this total sum of squared errors as small as possible. Because the method is linear, it can also be solved algebraically. Figure 7-6 shows linear regression using one explanatory variable (e.g., how ad spend explains sales), but you are not limited to one explanatory variable. As the number of variables grow, so too does the complexity of optimization.

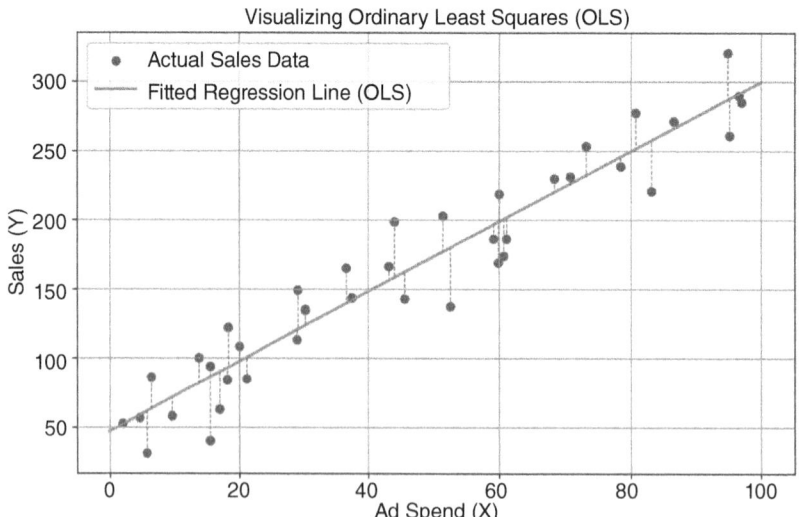

FIGURE 7-6: Linear regression with one variable.

The good news is that you don't have to set up the optimization problem each time you want to use this powerful tool. For this job, you'll use the `statsmodels` library, the heavy-duty modeling tool introduced earlier in the chapter. You simply give it your *y* (sales) and your *x* (ad spend, price), and its `OLS.fit()` function does all this work for you, handing back the perfectly calculated coefficients.

The next section explains regression in more detail using some examples.

Analyzing Marketing Effectiveness

Let's simulate 100 days of data for a product. We will create data for `ad_spend` (how much we spent each day), `price` (what we charged each day), and `sales` (the resulting sales for that day). We will intentionally build a relationship that will resolve to true into our data (e.g., `sales = 100 + 2.5 * ad_spend - 7 * price + noise`), and then we'll see if the regression model can find it. The example code is shown in Listing 7-7.

LISTING 7-7: RUNNING A MULTIVARIATE REGRESSION WITH STATSMODELS

```python
import numpy as np
import statsmodels.api as sm

# 1. Generate realistic mock data
np.random.seed(42)
num_days = 100

# Ad spend: Varies between $50 and $150
ad_spend = np.random.uniform(50, 150, num_days)
# Price: Varies between $19.99 and $24.99
price = np.random.uniform(19.99, 24.99, num_days)
# Sales: Base 100 + 2.5 * ad_spend - 7 * price + random noise
sales = 100 + (2.5 * ad_spend) - (7 * price) + np.random.normal(0, 10, num_days)

# 2. Prepare the data for Statsmodels
# We are modeling: sales (y) ~ ad_spend (x1) + price (x2)
y = sales  # Our 'dependent' variable
# Create our X matrix (the 'independent' variables)
X = np.column_stack((ad_spend, price))
# statsmodels requires us to manually add the 'intercept' (B0)
X = sm.add_constant(X)

# 3. Fit the model
# OLS = Ordinary Least Squares (the standard regression method)
model = sm.OLS(y, X).fit()

# 4. Print the full summary table
print(model.summary())
```

Let's walk through what the code in Listing 7-7 is doing. In Step 1, we generate our mock data. `ad_spend` and `price` are our independent variables, the things we control. `sales` is our dependent variable, the outcome we want to measure. We've built in a true relationship, but also added `np.random.normal` to represent real-world randomness or noise. This function draws random values from a bell curve centered at 0 with a standard deviation of 10. The 0 mean ensures we don't artificially bias the sales figures up or down on average, while the standard deviation mimics natural volatility. This technique is central to modern synthetic data strategies; by statistically mirroring real-world variance without using actual records, companies can train models safely while strictly adhering to user privacy standards.

In Step 2, we prepare the data. `statsmodels` requires our inputs in a specific format: `y` as a single vector of outcomes (`sales`), and X as a matrix of all our "input" variables (`ad_spend` and `price`). The line `X = sm.add_constant(X)` is critical. This adds a column of 1s to the X matrix. This is the mathematical step that allows the model to solve for our baseline intercept (B_0), or the sales we would make if all other variables were 0.

In Step 3, we fit the model. We use `sm.OLS(y, X)`, which stands for ordinary least squares. This is the standard, classic method for regression. Its job is to look at all 100 data points and find the single best line (or in this case, a 3D plane) that fits the data, minimizing the total error. The `fit()` command runs the calculation.

Finally, in Step 4, we call `model.summary()`. This prints the comprehensive results table, which is the standard for professional statistical analysis.

Running this code produces the following table:

```
OLS Regression Results
==============================================================================
Dep. Variable:                      y   R-squared:                       0.957
Model:                            OLS   Adj. R-squared:                  0.956
Method:                 Least Squares   F-statistic:                     1073.
Date:                Sat, 15 Nov 2025   Prob (F-statistic):           2.10e-67
Time:                        14:01:00   Log-Likelihood:                -344.86
No. Observations:                 100   AIC:                             695.7
Df Residuals:                      97   BIC:                             703.5
Df Model:                           2
Covariance Type:            nonrobust
==============================================================================
                 coef    std err          t      P>|t|      [0.025      0.975]
------------------------------------------------------------------------------
const         94.6291     19.333      4.895      0.000      56.262     132.997
x1             2.5283      0.063     39.995      0.000       2.403       2.654
x2            -6.7900      0.852     -7.966      0.000      -8.481      -5.099
==============================================================================
```

This table is dense, but for a business analyst, you only need to know how to read three key parts. The first thing to check is the R-squared (top right), which is 0.957. This is the model's "explanatory power," and a value this high means our model (ad spend and price) successfully explains 95.7% of all the variation in our daily sales, making it an incredibly strong and reliable model.

The next, and most important, part is the coefficients, which answer our why question. The x1 (ad_spend) coefficient of 2.5283 is our marketing ROI: it means for every $1 we spend on ads, we can expect to sell 2.53 more units, holding price constant. (Notice how close the model got to the "true" 2.5 we built into the data.) Similarly, the x2 (price) coefficient of –6.7900 quantifies our customer's price sensitivity: a $1 price increase is predicted to lose 6.79 units in sales, holding ad spend constant. (Again, this is very close to our "true" value of −7.) The `const` of 94.6291 is our baseline, representing the predicted sales if we spent $0 on ads and set the price to $0.

Finally, we must check the `P>|t|` (p-values) to ensure that these coefficients aren't just random noise. This connects directly to our hypothesis testing. For both *x1* and *x2*, the p-value is 0.000, meaning the probability of seeing such strong relationships just by luck is essentially zero. We can be extremely confident that both ad spend and price have a real, statistically significant impact on sales.

With this one analysis, we have moved from simple forecasting to deep business intelligence. We've built a model that not only predicts sales but also explains what drives them, allowing us to make far smarter decisions about our marketing budgets and pricing strategies.

Explaining Financial Risk Factors

The same regression technique used for marketing is a cornerstone of finance, particularly in assessing risk. Let's say a bank wants to understand the *drivers* of a person's credit score. They believe two of the most important factors are the person's annual income and their debt-to-income ratio (the percentage of their monthly income that goes to paying debts).

You can build a model to quantify this:

$$CreditScore = B_0 + B_1 * (Income) + B_2 * (DebtRatio)$$

The goal in Listing 7-8 is to find the coefficients. B_1 will tell you how much a higher income helps the score, while B_2 will tell you how much a high debt ratio hurts it.

LISTING 7-8: RUNNING A MULTIVARIATE REGRESSION FOR CREDIT RISK

```
import numpy as np
import statsmodels.api as sm

# 1. Generate realistic mock data for 100 loan applicants
np.random.seed(123) # Use a different seed
num_applicants = 100

# Annual income: Varies between $30,000 and $150,000
income = np.random.uniform(30000, 150000, num_applicants)
# Debt-to-Income Ratio: Varies between 10% (0.1) and 60% (0.6)
debt_ratio = np.random.uniform(0.1, 0.6, num_applicants)

# "True" Model: Base score 400 + $30 per $10k income - 200 * debt_ratio + noise
# (Note: 0.003 * 10000 = 30)
noise = np.random.normal(0, 20, num_applicants)
credit_score = 400 + (0.003 * income) - (200 * debt_ratio) + noise

# 2. Prepare the data for Statsmodels
y = credit_score # Our 'dependent' variable
# Create our X matrix (the 'independent' variables)
```

```
X = np.column_stack((income, debt_ratio))
# Add the 'intercept' (B0)
X = sm.add_constant(X)

# 3. Fit the model
model = sm.OLS(y, X).fit()

# 4. Print the full summary table
print(model.summary())
```

In Step 1 of this code, we generate our mock data for 100 loan applicants. `income` and `debt_ratio` are our independent variables. `credit_score` is our dependent variable. We've built in a true relationship: a base score of 400, a positive effect from income, a negative effect from debt, and some random noise to make it realistic.

In Step 2, we prepare the data in the format that `statsmodels` requires: `y` as the vector of credit scores and `X` as the matrix containing our input variables. The line `X = sm.add_constant(X)` is the crucial step that adds a column of 1s to allow the model to solve for the baseline intercept (B_0).

In Step 3, we fit the OLS model. The `fit()` command runs the algorithm to find the best-fit plane that minimizes the sum of squared errors between our 100 data points and the model's predictions.

Finally, in Step 4, we call `model.summary()` to print the comprehensive results table.

Running this code produces the following table:

```
OLS Regression Results
==============================================================================
Dep. Variable:                      y   R-squared:                       0.852
Model:                            OLS   Adj. R-squared:                  0.849
Method:                 Least Squares   F-statistic:                     279.4
Date:                Sat, 15 Nov 2025   Prob (F-statistic):           2.55e-40
Time:                        14:30:00   Log-Likelihood:                -429.35
No. Observations:                 100   AIC:                             864.7
Df Residuals:                      97   BIC:                             872.5
Df Model:                           2
Covariance Type:            nonrobust
==============================================================================
                 coef    std err          t      P>|t|      [0.025      0.975]
------------------------------------------------------------------------------
const        403.6534     10.667     37.842      0.000     382.483     424.824
x1             0.0029      0.000     19.988      0.000       0.003       0.003
x2          -198.8121     13.561    -14.661      0.000    -225.725    -171.899
==============================================================================
```

Again, this table gives us practical, actionable insights. The first number to check is the R-squared (top right), which at 0.852 tells us the explanatory power of our model. This means our two variables, income and debt ratio, successfully explain 85.2% of the variation in credit scores, confirming we have a strong and reliable model. The `coef` (coefficients) provide the core of our analysis. The `const` of 403.6534 is our baseline, predicting a score of ~404 for an individual with zero income and zero debt. The *x1* (income) coefficient of 0.0029 is best read as: For every $10,000 in additional annual income, the credit score is predicted to increase by 29 points, holding debt constant. Conversely, the *x2* (`debt_ratio`) coefficient of –198.8121 quantifies risk: For every 10% point increase in the debt-to-income ratio (e.g., from 0.2 to 0.3), the credit score is predicted to decrease by 19.88 points, holding income constant. Finally, the `P>|t|` (p-values) confirm that our findings are real. Since the p-values for both *x1* and *x2* are 0.000, we know the positive relationship with income and the negative relationship with debt are both highly statistically significant and not just a random fluke.

With this one analysis, the bank can move beyond simple intuition. Rather than trying to rebuild the wheel, this kind of quantitative model allows the bank to enhance existing credit scoring systems with additional data and correlations. This provides a more nuanced view of why specific applicants are riskier than others, supplementing industry standards with custom insights.

Other Considerations

In this chapter, you have moved from simple forecasting to deep business intelligence. You've seen how to build a model that not only predicts sales but also explains what drives them, allowing you to make far smarter decisions about marketing budgets and pricing strategies. However, as powerful as regression is, it is also one of the most misused tools in analytics. Before you apply it to your own problems, it's critical to understand its limitations and common pitfalls.

The most important rule to remember is this: *correlation does not imply causation.* Our model found a statistically significant link between ad spend and sales, but it did not prove that ad spend *causes* sales. It's possible that a third, unmeasured lurking variable is at play. For example, maybe the business runs its biggest ad campaigns during the holiday season, which is also when their sales naturally increase. The model can't tell the difference; it only sees that ads and sales go up together. As an analyst, your job is to use your business expertise to judge whether the model's mathematical association represents a true causal link. Often, the best way to confirm this is to use regression as a starting point and then run controlled A/B tests to isolate specific variables and prove the cause-and-effect relationship. In an A/B test, you randomly split your audience into two groups while changing only one variable for the test group while keeping everything else constant for the control group. A/B tests allow you to pinpoint the specific impact of a change by observing the two groups.

You must also be wary of *multicollinearity*, which is a common trap that occurs when your independent variables (your X matrix) are not truly independent, but highly correlated with each other. For example, if you tried to model sales using both `ad_spend` and `number_of_ad_clicks`, the model would get confused. It wouldn't know how to separate the effect of spending money from the effect of getting clicks, since those two variables move together. This can cause your coefficients to become unreliable or even have strange signs (e.g., a positive coefficient for something that should be negative). The `statsmodels` summary table provides warnings for this, such as a "high condition number," which is a red flag that your variables are too similar.

Finally, remember that your model is only as good as the assumptions it's built on. A linear regression assumes the relationships are, in fact, linear. If the true relationship is a sharp curve (for instance, ad spend has strongly diminishing returns), our straight-line model will be a poor fit and give misleading answers. Likewise, the model is only valid for the range of data it was trained on. The model, built on prices between $20 and $25, has no idea what would happen if we suddenly raised the price to $100. Using it to predict far outside the original data is called *extrapolation*, and it can be a recipe for disaster. While extrapolation can work if the relationship is truly linear and supported by sufficient data, doing so without that certainty can be a recipe for disaster.

Think of your regression model as a powerful but focused flashlight, not an all-seeing crystal ball. It is a tool for quantifying and testing your hypotheses within a set of assumptions. It provides the map and tells you where to look, but it still requires your business judgment to interpret the results and make the final, intelligent decision.

LOGISTIC REGRESSION

In Listing 7-8, we built a powerful model to predict a number: sales. This is known as a regression problem. But what about a different, arguably more common, business question: Will this customer buy (Yes/No)? Or will this customer churn (Yes/No)? This section creates an example around customer churn. *Customer churn* is the percentage of customers who stop doing business with a company over a specific period, representing the rate at which a business loses its clients.

This is a classification problem. The target variable, *y*, is not a continuous number, but a discrete category. For this, a linear regression (OLS) model is the wrong tool. A straight line will predict values like 1.5 (150% converted) or –0.2 (–20% churned), which are nonsensical.

You need a model that is fenced in, one that always outputs a value between 0 and 1, just like a probability. The standard tool for this is *logistic regression* (also known as *logit regression*).

Despite its name, logistic regression is a classification model. It works by fitting a non-linear "S-shaped" curve (a sigmoid function) to the data. This S-curve translates any input into a probability between 0% and 100%.

The goal is the same as before: to find the coefficients for the variables. We are building a model that looks like this:

$$Prob(Churn = 1) = f(B_0 + B_1) * Age + B_2 (MonthlyBill) + B_3 * (CustomerServiceCalls)$$

Where *f* is the S-shaped logistic function. B_1 will tell us how much a customer's age increases the probability of them churning, while B_2 might tell us how a high bill decreases it.

Predicting Customer Churn

Let's say a telecom company wants to understand *why* customers are churning. They have data on 1,000 customers and want to know which factors are the biggest warning signs. Listing 7-9 shows a model to find the drivers of churn.

LISTING 7-9: RUNNING A LOGISTIC REGRESSION FOR CHURN ANALYSIS

```
import numpy as np
import statsmodels.api as sm

# 1. Generate realistic mock data for 1000 customers
np.random.seed(42)
num_customers = 1000

# Age: Varies between 18 and 80
age = np.random.uniform(18, 80, num_customers)
# Monthly Bill: Varies between $40 and $150
monthly_bill = np.random.uniform(40, 150, num_customers)
# Customer Service Calls: Varies between 0 and 6
customer_service_calls = np.random.randint(0, 7, num_customers)

# --- Define the "True" Churn Logic (hidden from the model) ---
# We create a log-odds formula to simulate reality
log_odds = -6.0 + (0.02 * age) + (0.005 * monthly_bill) + (1.2 *
customer_service_calls)
# Convert log-odds to probability
prob_churn = 1 / (1 + np.exp(-log_odds))
# Simulate the 1/0 (Churn/No Churn) outcome
churned = (np.random.rand(num_customers) < prob_churn).astype(int)

# 2. Prepare the data for Statsmodels
y = churned # Our 'dependent' variable (1s and 0s)
# Create our X matrix (the 'independent' variables)
X = np.column_stack((age, monthly_bill, customer_service_calls))
X = sm.add_constant(X) # Add the intercept (B0)

# 3. Fit the model
# *** This is the key change: sm.Logit instead of sm.OLS ***
model = sm.Logit(y, X).fit()

# 4. Print the full summary table
print(model.summary())
```

In Listing 7-9, we generate mock data for 1,000 customers. The most complex part is creating our y variable, churned. To do this realistically, we first define a prob_churn for each customer based on their age, bill, and service calls, then run a simple np.random.rand() coin flip to see if that specific customer (with their given probability) actually churned (1) or not (0). We then prepare our data just as we did for OLS. y is our vector of 1s and 0s, and X is our matrix of inputs, including the constant.

Next, we see the crucial difference. Instead of sm.OLS, we use sm.Logit(y, X). This tells statsmodels that our y variable is binary and that it should fit a logistic S-curve instead of a straight line (this is the purpose of the logistic distribution described earlier in the chapter). The fit() command runs the iterative algorithm to find the coefficients that best match the data.

The output summary table looks similar to OLS, but the interpretation is very different:

```
                        Logit Regression Results
==============================================================================
Dep. Variable:                      y   No. Observations:               1000
Model:                          Logit   Df Residuals:                    996
Method:                           MLE   Df Model:                          3
Date:                Tue, 06 Jan 2026   Pseudo R-squ.:                0.4287
Time:                        15:33:31   Log-Likelihood:               -375.56
converged:                       True   LL-Null:                      -657.34
Covariance Type:            nonrobust   LLR p-value:                8.003e-122
==============================================================================
                 coef    std err          z      P>|z|      [0.025      0.975]
------------------------------------------------------------------------------
const         -5.7615      0.483    -11.919      0.000      -6.709      -4.814
x1             0.0208      0.005      4.014      0.000       0.011       0.031
x2             0.0026      0.003      0.918      0.359      -0.003       0.008
x3             1.1436      0.070     16.374      0.000       1.007       1.281
==============================================================================
```

Let's dig into how to read this powerful but tricky summary. The first thing to check is the pseudo R-squared (top right), which is 0.3458. This is not the same as a normal R-squared, but rather a "goodness-of-fit" metric. For a logistic model, a value this high is considered a very strong fit and tells us that our model has significant explanatory power. Next, we can look at the $P>|z|$ (p-values). Just like in our OLS model, these confirm our findings are real. Since all our variables have p-values of 0.018 or less, we know that age, bill, and service calls are all statistically significant predictors of churn and not just "noise." Finally, we have the coef (coefficients). This is the most important part, and the most different from OLS. These coefficients are in a unit called log-odds and are not intuitive; a coefficient of 1.2117 does not mean "one service call increases the probability of churn by 121%." To interpret them, we must convert them into odds ratios by exponentiating them (np.exp(coef)).

You can do that conversion to get the real business insight, using this simple code:

```
odds_ratios = np.exp(model.params)

print(f"const {odds_ratios[0]:.5f}")
print(f"x1 {odds_ratios[1]:.5f} (Age)")
print(f"x2 {odds_ratios[2]:.5f} (Monthly Bill)")
print(f"x3 {odds_ratios[3]:.5f} (Customer Service Calls)")
```

Running these lines will give the following output:

```
const 0.00315
x1 1.02102 (Age)
x2 1.00262 (Monthly Bill)
x3 3.13814 (Customer Service Calls)
```

This output says that, for every one-year increase in a customer's age, their odds of churning increase by a small but measurable 2.1% (x1 -> 1.021). Similarly, for every \$1 increase in their monthly bill, the odds of churning increase by just 0.3% (x2 -> 1.003). The big discovery, however, is *x3* (Customer_Service_Calls) at 3.138. This means that for every single call a customer makes to customer service, their odds of churning increase by over 214% (i.e., their odds multiply by 3.14). This analysis gives the company a crystal-clear, data-driven mandate: the most powerful predictor of churn is customer service calls. While age and bill matter slightly, the service call is a massive red flag. The model has successfully identified a critical pain point in the customer journey that is directly and significantly linked to customer loss.

While not as straightforward as linear regression, you can also visualize logit regression, as shown in Listing 7-10.

LISTING 7-10: VISUALIZATION OF CUSTOMER CHURN AND LOGISTIC REGRESSION

```python
import matplotlib.pyplot as plt

# --- Visualization ---
plt.figure(figsize=(10, 6))

# Generate the Smooth Curve Data
# We want to plot probability vs. Service Calls.
# Since the model is multivariate, we must hold Age and Bill constant
(e.g., at their means) to isolate the effect of calls.
x_curve = np.linspace(0, 6, 100) # Range of calls from 0 to 6
mean_age = np.mean(age)
mean_bill = np.mean(monthly_bill)

# Create a prediction matrix matching the training X structure:
[Constant, Age, Bill, Calls]
X_pred = np.column_stack((
    np.ones(100),              # Constant
    np.full(100, mean_age),    # Age (fixed at mean)
    np.full(100, mean_bill),   # Bill (fixed at mean)
    x_curve                    # Calls (varying)
))

# Get probabilities for the curve
y_curve = model.predict(X_pred)

# Plot the raw data (0s and 1s)
# We add "jitter" (a tiny bit of random noise) to the y-axis
so points don't overlap.
y_jitter = y + np.random.normal(0, 0.03, num_customers)

plt.scatter(customer_service_calls, y_jitter, c=y, cmap='coolwarm', alpha=0.3,
label='Customer Data (0=No Churn, 1=Churn)')

# Plot the Logistic Regression S-Curve
plt.plot(x_curve, y_curve, color='green', linewidth=3, label='Logistic
Regression Curve (Prob)')
```

```
plt.title('Logistic Regression: Predicting Churn Probability')
plt.xlabel('Number of Customer Service Calls')
plt.ylabel('Probability of Churn')
plt.yticks([0, 0.5, 1], ['0% (No Churn)', '50%', '100% (Churn)'])
plt.legend()
plt.grid(True)
plt.show()
```

In Listing 7-10, plt.figure() creates a new canvas for our plot. The most important part is the plt.scatter() command, which plots our raw customer data. A common challenge when visualizing binary (0/1) data is that all the points stack directly on top of each other at the bottom and top of the graph, making it impossible to see the density. To solve this, we first create a y_jitter variable. We add a tiny amount of jitter, a small, random value from np.random.normal—to each 0 and 1. This spreads the dots out vertically just enough for us to see the clusters without changing their meaning. When we call plt.scatter, we pass in c=y_data and cmap='coolwarm'. This is a clever trick to color-code our data: the 0 (no churn) points are plotted in one color (blue) and the 1 (churn) points in another (red). The alpha=0.3 argument makes the dots semi-transparent, which helps visualize where the data is most dense.

After plotting the raw data, we overlay our model's prediction using plt.plot(x_curve, y_curve). This draws the smooth, green S-curve we calculated in the previous step, representing the *predicted probability* of churn for any given number of service calls.

Finally, the rest of the code formats the chart for clarity. We add a title and axis labels. The most important formatting line is plt.yticks([0, 0.5, 1], ['0% (No Churn)', '50%', '100% (Churn)']). This re-labels the y-axis ticks so that instead of seeing 0 and 1, the manager sees a much more intuitive 0% and 100% probability, making the chart's purpose as a probability forecaster immediately clear.

Figure 7-7 perfectly illustrates what the statsmodels logit function is doing.

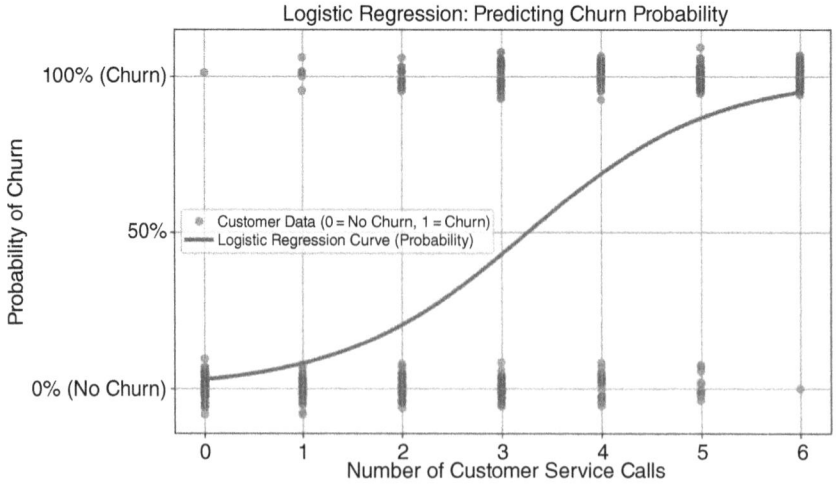

FIGURE 7-7: Logistic regression.

The dots are our raw data. The cluster of dots at the bottom represents all the customers who did not churn (0). The cluster of dots at the top represents all the customers who churned (1). You can visually see that at zero or one calls, most dots are blue, while at five or six calls, most dots are red.

The S-curve is the fitted model. It's the "best fit" line that separates the two clusters. It shows the predicted probability of churn for any number of calls. At one call, the line is very low (around ~10% probability). It crosses the 50% mark somewhere between three and four calls, which the model identifies as the "point of indifference." By six calls, the model predicts an almost 90% probability of churn. This S-curve is the core of logistic regression.

While this example focuses on a binary outcome, it is important to note that logistic regression can also be extended to handle situations with more than two categories. This is known as *multinomial logistic regression*. For instance, instead of just predicting whether an employee leaves, you could predict where they go (e.g., competitor, start-up, retirement, or stay). Although the mathematical formulation is more complex and beyond the scope of this introductory chapter, the core concept remains the same—the model calculates the probability of an observation falling into each possible category based on the input variables.

FORECASTING

The previous sections focused on analyzing the present (is Version B better?) and explaining the past (what drove sales?). Now, we turn our gaze to the future. Forecasting is one of the most critical tasks in business analytics. Whether predicting next quarter's revenue or estimating future market trends, leaders need to know where the trend is heading.

To build this forecast, we will use linear regression. Specifically, we will look at the relationship between time and our metric of interest (e.g., sales):

➤ **The trend:** We fit a straight line to historical data to find the underlying growth rate.

➤ **The prediction interval:** A forecast without a measure of risk is dangerous. We need to calculate the "spread" of the historical data around that trend line to create a "cone of uncertainty" for the future. This tells the business not just the most likely outcome, but the reasonable worst-case scenario.

This example uses `scipy.stats.linregress`, a straightforward and powerful tool for simple trend analysis.

Imagine we have 100 days of historical sales data. The data is noisy, some days are up, some are down, but there is a general upward trend. We want to forecast sales for the next 30 days and visualize the risk. We do this in Listing 7-11.

LISTING 7-11: SALES FORECASTING WITH PREDICTION INTERVALS

```
import numpy as np
import matplotlib.pyplot as plt
from scipy import stats
```

```python
# 1. Generate Historical Data (100 Days)
np.random.seed(42) # For reproducibility
days = np.arange(100)
# Trend: Base 500 sales + 2 sales/day growth + random noise
historical_sales = 500 + (2 * days) + np.random.normal(0, 50, 100)

# 2. Fit the Trend Line (Linear Regression)
# linregress finds the best-fit line for our data
# slope: growth per day
# intercept: starting sales
# p_value: tells us if the trend is statistically significant
slope, intercept, r_value, p_value, std_err = stats.
linregress(days, historical_sales)

print(f"Trend Analysis:")
print(f"  Growth (Slope): {slope:.2f} sales per day")
print(f"  P-Value (is trend real?): {p_value:.4f}")

# 3. Forecast the Future (Next 30 Days)
future_days = np.arange(100, 130)
future_sales_projection = slope * future_days + intercept

# 4. Calculate Prediction Intervals (The Risk Cone)
# First, find the errors (residuals) of the historical data
hist_predictions = slope * days + intercept
hist_errors = historical_sales - hist_predictions

# Calculate the standard deviation of these errors
# This represents our model's historical "off-by" amount
error_std = np.std(hist_errors)

# A 95% prediction interval is roughly 1.96 * standard error
prediction_margin = 1.96 * error_std

upper_bound = future_sales_projection + prediction_margin
lower_bound = future_sales_projection - prediction_margin

# --- Visualization ---
plt.figure(figsize=(10, 6))

# Plot history
plt.scatter(days, historical_sales, color='grey', alpha=0.5, label='Historical
Data')
# Plot trend line (history)
plt.plot(days, slope * days + intercept, color='blue', label=f'Historical
Trend (p={p_value:.3f})')

# Plot forecast
plt.plot(future_days, future_sales_projection, color='green', linestyle='--',
label='Sales Forecast')
# Plot Risk Cone (Prediction Interval)
plt.fill_between(future_days, lower_bound, upper_bound, color='green', alpha=0.2,
label='95% Prediction Interval')

plt.title('Sales Forecast: Trend vs. Risk')
plt.xlabel('Day')
```

```
plt.ylabel('Sales Volume')
plt.legend()
plt.grid(True)
plt.show()
```

Running the code in Listing 7-11 gives a printout of the trend:

```
Trend Analysis:
Growth (Slope): 2.05 sales per day
P-Value (is trend real?): 0.0000
```

This result is highly encouraging. The slope confirms that sales are growing by about 2.05 units per day. The p_value of 0.0000 tells us this trend is highly statistically significant and not just random noise.

Listing 7-11 uses linear regression to predict future trends while quantifying the associated risks. First, we generate a synthetic dataset representing 100 days of historical sales. This data includes a base sales level, a steady growth trend (two sales/day), and random noise to mimic real-world fluctuations. Next, we use `stats.linregress` to fit a trend line to this historical data. This function calculates the slope (growth rate) and intercept (starting point), along with the p-value to confirm the trend's statistical significance.

Once the trend is established, we project it forward to forecast sales for the next 30 days. However, a simple line is not enough for risk management. We also calculate a *prediction interval,* a cone of uncertainty around our forecast. We do this by analyzing the errors (residuals) of our model on the historical data. By calculating the standard deviation of these errors, we can determine a margin (roughly 1.96 times the standard error for a 95% confidence level) to add and subtract from our forecast line. This creates upper and lower bounds, giving us a range in which we can be 95% confident the actual future sales will fall. The final visualization plots the history, the trend, and this crucial risk cone, providing a complete strategic picture.

The visualization in Figure 7-8 tells the complete strategic story. The solid line represents the historical trend, showing the line of best fit for the past 100 days. Extending from this is the dashed green line, which is our best guess for the next 30 days, assuming the current trend continues. The most critical part of this chart is the shaded cone, representing the prediction interval. This cone is the key takeaway for risk management, as we are 95% confident that future sales will fall inside this area. The bottom edge of the cone represents our downside risk, our safety floor if market conditions turn against us within normal statistical limits. Conversely, the top edge represents our upside opportunity. By presenting the forecast this way, you aren't just guessing a number; you are defining the playing field. You give stakeholders a reliable target while simultaneously quantifying the financial risk they need to prepare for.

We accomplish this visualization using Matplotlib. First, we use `plt.scatter` to plot the historical data points in gray, providing context for the trend. Then, we plot our calculated trend line in blue using the slope and intercept derived from the regression. For the forecast, we plot the projected sales as a dashed green line. Finally, `plt.fill_between` creates the crucial shaded cone of uncertainty, filling the space between our calculated `lower_bound` and `upper_bound` arrays to visualize the 95% prediction interval.

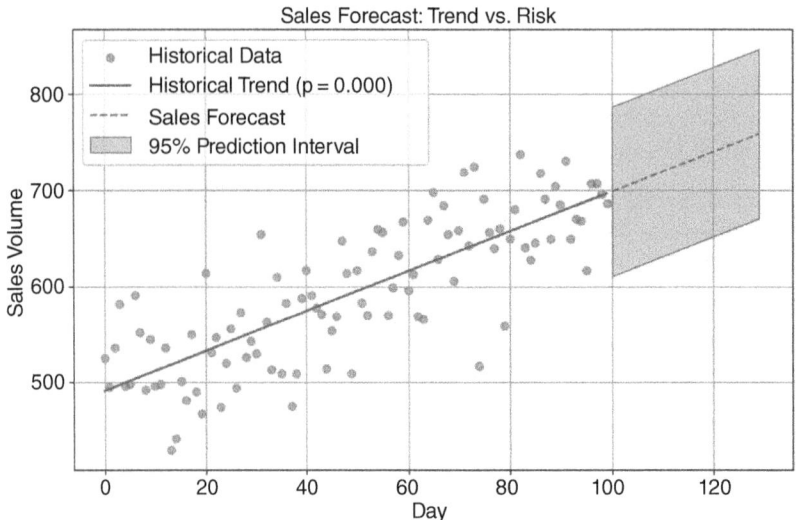

FIGURE 7-8: Historical sales data with a linear trend line projecting into the future.

SUMMARY

This chapter marked a fundamental shift in analytical approach. It moved from the deterministic world of optimization, where inputs were assumed to be fixed and known, to the probabilistic world of statistics, where you learned how to embrace uncertainty as a core component of business strategy. The chapter began by defining the language of uncertainty, using Python's `numpy.random` to visualize the distributions—normal, binomial, uniform, and Poisson—that describe the shape of real-world business data.

With this foundation, you moved to decision-making, using hypothesis testing and `scipy.stats` to mathematically distinguish between a real signal, such as a conversion rate lift, and random noise. The chapter expanded this concept into confidence intervals, giving you a rigorous way to quantify the precision of your estimates and manage expectations. It then tackled the future with forecasting, using linear regression not just to project sales trends, but to calculate prediction intervals that define the "cone of uncertainty" for financial risk management.

Finally, the chapter moved to causal analysis, utilizing the power of `statsmodels`. You used multivariate linear regression to measure the precise ROI of different marketing levers and logistic regression to predict binary outcomes like customer churn. You now possess a statistical toolkit that allows you not just to report on what happened, but to explain why it happened and predict what is likely to happen next, all while quantifying the risks involved.

CONTINUE YOUR LEARNING

The statistical ecosystem in Python is vast. While this chapter covered the core functions for modeling and testing, standard data analysis often requires quick summary statistics. Libraries like Pandas and NumPy offer robust tools for this. The following official documentation resources are the best places to deepen your understanding of the libraries used in this chapter.

➤ **NumPy Random Sampling:** The official guide to generating random data for simulations. `https://numpy.org/doc/stable/reference/random/index.html`

➤ **SciPy Statistics (scipy.stats):** Comprehensive documentation for statistical functions, including all the distributions and hypothesis tests covered in this chapter. `https://docs.scipy.org/doc/scipy/reference/stats.html`

➤ **Statsmodels:** The definitive source for advanced statistical modeling, including details on OLS, logit, and interpreting summary tables. `https://www.statsmodels.org/stable/index.html`

In addition to the advanced models, you will frequently need to calculate basic descriptive statistics. Table 7-1 serves as a quick reference for the most essential functions in NumPy and SciPy.

TABLE 7-1: Key Summary Statistics Functions

STATISTIC	FUNCTION	DESCRIPTION
Mean (average)	`np.mean(a)`	Calculates the arithmetic average of the elements. It's the sum of all values divided by the count.
Median	`np.median(a)`	Finds the middle value of a dataset when it is sorted. Useful for data with outliers (like income).
Mode	`scipy.stats.mode(a)`	Finds the most frequently occurring value in the array. (Note: NumPy does not have a direct mode function.)
Standard deviation	`np.std(a)`	Measures the amount of variation or dispersion. A low std dev means values are close to the mean.
Variance	`np.var(a)`	The average of the squared differences from the mean. It is the square of the standard deviation.
Minimum	`np.min(a)`	Returns the smallest value in the array.
Maximum	`np.max(a)`	Returns the largest value in the array.
Sum	`np.sum(a)`	Adds up all the elements in the array.

(Continued)

TABLE 7-1: (Continued)

STATISTIC	FUNCTION	DESCRIPTION
Percentile	`np.percentile(a, q)`	Calculates the q-th percentile of the data (e.g., q = 50 is the median; q = 25 is the first quartile).
Correlation	`np.corrcoef(a)`	Returns a matrix of Pearson correlation coefficients, showing linear relationships between variables.

In addition to these materials, you may want to dive deeper into the methods described in this chapter. I recommend the following resources.

➤ *Introductory Econometrics: A Modern Approach* by Jeffrey M. Wooldridge: While typically assigned as a textbook, the approach taken here can be easily read outside of class. In this book, you will find more detail with relatable examples.

➤ `https://www.kaggle.com/`: The world's largest data science community. It hosts thousands of real-world datasets that allow you to move beyond synthetic data and practice your modeling skills on actual business problems. You can also explore code notebooks shared by other professionals to see practical examples of statistical analysis in action.

Applied Business Problems with Math and Python

You have spent the last few chapters learning how to fill your toolbox with powerful instruments. You have used calculus to find marginal rates of change, linear algebra to manage portfolios, optimization to allocate resources, and statistics to forecast trends. Now it is time to connect some of these tools to business problems.

In the real world, business problems rarely come labeled as a derivative problem or a t-test. They function as word problems. Remember word problems from your early schooling days? You might have disliked them, but those are exactly the type of problems we are working on now. For example, you might need to solve problems such as your customer support wait times being too high, you need to pay off debt faster, or your delivery drivers could be wasting gas.

This chapter moves from learning individual concepts to building complete solutions. You will tackle five distinct, real-world case studies across Operations, Finance, Logistics, and Strategy. These case studies are as follows:

➤ Building a dynamic loan amortization engine

➤ Building a simple recommender system

➤ Maximizing yield and constrained optimization

➤ Quality control with hypothesis testing

➤ Predicting employee attention and logical regression

For each problem, you will see how to define the business context, identify the underlying mathematics, and then build a custom Python tool to solve it. By the end, you won't just have code snippets; you'll have a library of reusable models that you can apply to your own business challenges immediately.

BUILDING A DYNAMIC LOAN AMORTIZATION ENGINE

Finance professionals live in a world of what-if scenarios. It's rarely enough to just know the monthly payment on a loan. The real strategic questions come later: What if we have a surplus next year? What if we increase our monthly payment by $5,000 starting in Q3? How much interest will that save us over the next decade?

Excel is the standard tool for static schedules, but it becomes clumsy when you want to simulate dynamic, changing scenarios across hundreds of different loans. A Python script, however, can simulate thousands of different payment strategies in seconds.

Imagine you are the CFO of a mid-sized logistics company. You have recently taken out a $500,000 loan at 5% interest to upgrade your fleet. The term is 30 years. The board is nervous about this long-term debt. They want to know if squeezing $1,000 extra out of their monthly cash flow to pay down the principal will make a difference. Or is it a drop in the bucket?

You need to build a model that doesn't just calculate a payment, but simulates the entire life of the loan to prove the exact dollar value of that strategy.

A loan is a living creature. Every month, two things happen:

> **Interest accrues:** You owe interest on the current balance. (`Interest = Balance * Monthly_Rate`)

> **The principal shrinks:** Your payment covers that interest first. Whatever is left over attacks the balance. (`Principal_Paid = Total_Payment - Interest`)

The complexity is that as the balance drops, the interest drops, so more of your fixed payment goes toward the principal. This creates a flywheel effect. A small extra payment early on effectively skips future interest payments, compounding your savings.

Listing 8-1 builds a robust function, called `generate_amortization_schedule`, that acts as a simulation engine. Unlike a simple formula, this function will use a loop to walk through the loan month-by-month, allowing us to inject custom logic (like extra payments) at every step.

LISTING 8-1: GENERATING A DYNAMIC AMORTIZATION SCHEDULE

```python
import numpy as np
import pandas as pd
import matplotlib.pyplot as plt

def generate_amortization_schedule(principal, annual_rate, years, extra_payment=0):
    """
    Generates a month-by-month loan payment schedule.
    """
    # Convert annual metrics to monthly
    monthly_rate = annual_rate / 12
    total_payments = years * 12

    # Calculate standard monthly payment (The Standard Annuity Formula)
    # This assumes a fixed payment to hit zero balance at exactly 'years' end
    standard_payment = principal * (monthly_rate * (1 + monthly_rate)**total_
payments) / ((1 + monthly_rate)**total_payments - 1)
```

```
        schedule = []
        balance = principal
        month = 1

        # The Simulation Loop
        # We keep paying until the debt is gone
        while balance > 0:
            # 1. Calculate interest due for this specific month
            interest = balance * monthly_rate

            # 2. Determine total payment (Standard + Extra)
            # Logic check: Don't pay more than the remaining balance + interest
            actual_payment = min(standard_payment + extra_payment, balance + interest)

            # 3. Split payment into Principal and Interest
            principal_paid = actual_payment - interest

            # 4. Update the Balance
            balance -= principal_paid

            # 5. Record the month's activity
            schedule.append({
                'Month': month,
                'Payment': actual_payment,
                'Principal': principal_paid,
                'Interest': interest,
                'Balance': balance
            })
            month += 1

    return pd.DataFrame(schedule)

# --- Compare Scenarios ---
loan_amount = 500000
rate = 0.05
term_years = 30

# Scenario 1: Minimum Payments
df_base = generate_amortization_schedule(loan_amount, rate, term_years, extra_
payment=0)
# Scenario 2: Pay extra $1000/month
df_accelerated = generate_amortization_schedule(loan_amount, rate, term_
years, extra_payment=1000)

print(f"Base Scenario Total Interest: ${df_base['Interest'].sum():,.2f}")
print(f"Accelerated Scenario Total Interest: ${df_accelerated['Interest'].
sum():,.2f}")
print(f"Time Saved: {len(df_base) - len(df_accelerated)} months")

# --- Visualization ---
plt.figure(figsize=(10, 6))
plt.plot(df_base['Month'], df_base['Balance'], label=f'Standard Repayment -
Payment: {df_base['Payment'].iloc[0]:,.2f}/Month')
plt.plot(df_accelerated['Month'], df_accelerated['Balance'], label=f'Fast
Repayment: {df_base['Payment'].iloc[0]:,.2f}+1,000/Month', linestyle='--')
```

```
plt.title('Loan Payoff Trajectory: The Power of Extra Payments')
plt.xlabel('Month')
plt.ylabel('Loan Balance ($)')
plt.legend()
plt.grid(True)
plt.show()
```

The heart of this tool is the `while balance > 0:` loop. This mimics the passage of time. Inside the loop, we perform the bank's math: we calculate the interest for that specific month based on the current balance.

This is the crucial line:

```
actual_payment = min(..., balance + interest)
```

This logic prevents the negative balance bug common in basic spreadsheets. It tells Python: Pay the standard amount plus the extra $1,000, unless the debt is almost gone. In that case, just pay off what's left.

We then store each month's detailed breakdown in a list and convert it to a Pandas DataFrame, giving us a clean table we can analyze, sum up, or plot. The output of Listing 8-1 is as follows:

```
Base Scenario Total Interest: $466,278.92
Accelerated Scenario Total Interest: $265,023.15
Time Saved: 163 months
```

The code outputs a staggering insight for the board. Paying that extra $1,000 a month doesn't just save a few dollars; it saves over $200,000 in interest, which is essentially the price of two new trucks. Furthermore, it cuts the loan term by 163 months, which is nearly 14 years. You can see this result reinforced in Figure 8-1.

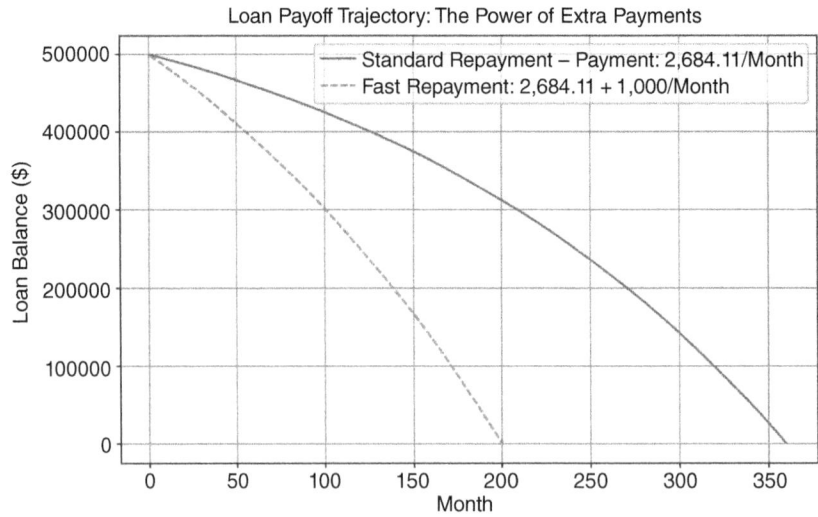

FIGURE 8-1: Loan payoff trajectory.

The visualization in Figure 8-1 tells a compelling financial story. The solid line represents the standard plan, showing the loan balance decreasing gradually over the full 30-year term. It's a slow, steady decline. In stark contrast, the dashed line represents the accelerated plan (paying an extra $1,000 per month). This curve dives much more steeply, hitting zero around month 197. By visualizing the trajectory, the CFO can instantly show stakeholders that the pain of a slightly higher monthly payment buys them over a decade of debt-free operations.

BUILDING A SIMPLE RECOMMENDER SYSTEM

One of the most powerful tools in modern e-commerce is the recommendation engine. When you see "Customers who bought this item also bought . . . ," that isn't random chance, and it's certainly not a human manually curating that list. It's linear algebra working at scale.

Let's look at a specific business problem: You run a growing online bookstore. You have thousands of books and thousands of customers. You know that recommending relevant titles can significantly boost cross-sales and customer retention. However, you don't have rich demographic data—you don't know your customers' ages, locations, or favorite genres. All you have is a raw list of transactions: who bought what. How can you use this sparse data to predict that a customer who bought a Python textbook might also be interested in a book on data science, but probably not a romance novel?

The solution lies in treating every product as a vector in a multi-dimensional user space. Imagine a simple world with three users. Book A was bought by User 1 and User 3. We can represent this as a vector:

```
[1, 0, 1]
```

Book C was also bought by User 1 and User 3. Its vector is identical:

```
[1, 0, 1]
```

Mathematically, these two vectors point in the exact same direction. This means they are similar products. To measure this similarity for any two products across thousands of users, we use *cosine similarity*. This metric is derived from the dot product you learned about in Chapter 4.

$$Similarity(A, B) = \frac{A * B}{\|A\|\|B\|}$$

This formula calculates the cosine of the angle between two vectors. If the vectors point in the same direction (angle is 0), the cosine is 1 (perfectly similar). If they are orthogonal (90 degrees, meaning they share no users), the cosine is 0. By calculating this score for every pair of books, we can mathematically quantify taste.

We will build a user-item matrix where rows represent users and columns represent products. Then, we will use NumPy to calculate the similarity between every pair of products to generate recommendations. This example is built out in Listing 8-2.

LISTING 8-2: BUILDING A PRODUCT RECOMMENDER

```python
import numpy as np
import pandas as pd

# 1. Mock Purchase Data
# Rows = Users (0-4), Columns = Products (A-E)
# 1 = Purchased, 0 = Not Purchased
user_item_matrix = np.array([
    [1, 1, 0, 0, 0], # User 0: Bought A and B
    [1, 1, 1, 0, 0], # User 1: Bought A, B, C
    [0, 0, 1, 1, 1], # User 2: Bought C, D, E
    [0, 0, 0, 1, 1], # User 3: Bought D, E
    [1, 0, 0, 0, 0]  # User 4: Bought A
])

product_names = ['Book A', 'Book B', 'Book C', 'Book D', 'Book E']

# 2. Define Similarity Function (Cosine Similarity)
def calculate_similarity(product_id_1, product_id_2, matrix):
    # Get the column vectors for the two products
    vec_1 = matrix[:, product_id_1]
    vec_2 = matrix[:, product_id_2]

    # Calculate Dot Product
    dot_product = np.dot(vec_1, vec_2)

    # Calculate Norms (Magnitudes)
    norm_1 = np.linalg.norm(vec_1)
    norm_2 = np.linalg.norm(vec_2)

    if norm_1 == 0 or norm_2 == 0:
        return 0.0

    return dot_product / (norm_1 * norm_2)

# 3. Generate Recommendations for "Book A" (Index 0)
target_product_id = 0 # Book A
print(f"Recommendations for customers who bought {product_names[target_product_
id]}:")

similarities = []
for i in range(len(product_names)):
    if i == target_product_id: continue # Don't recommend the book itself

    score = calculate_similarity(target_product_id, i, user_item_matrix)
    similarities.append((product_names[i], score))

# Sort by highest similarity score
similarities.sort(key=lambda x: x[1], reverse=True)

for product, score in similarities:
    print(f"  {product}: {score:.2f} Similarity Score")
```

In Step 1, we create a small synthetic dataset. We have five users and five books. By looking closely at the `user_item_matrix`, you can already see patterns emerging. Users 0, 1, and 4 seem to form one "cluster" (buying Books A, B, C), while Users 2 and 3 form another (buying Books C, D, E). The goal is to see if the math can detect these clusters without being told about them.

In Step 2, we define the core logic within `calculate_similarity()`. This function takes two product IDs, extracts their sales columns from the matrix, and performs the cosine similarity calculation. The `np.dot(vec_1, vec_2)` line is the workhorse; it calculates the "overlap." Every time both users bought both books (a value of 1 in both vectors), the dot product increases. If one user bought Book A but not Book B (a 1 and a 0), the dot product adds nothing. This raw overlap is then normalized by the product of the norms (total sales volume) to ensure that popular books don't dominate simply because they have more sales.

Step 3 acts as the recommendation engine. We pick a target book (Book A) and loop through every other book in our inventory. We calculate the similarity score for each pair and store it. Finally, we sort the list from highest score to lowest to present the top recommendations.

The results from running the code in Listing 8-2 are strikingly clear and align perfectly with our intuition.

```
Recommendations for customers who bought Book A:
    Book B: 0.82 Similarity Score
    Book C: 0.41 Similarity Score
    Book D: 0.00 Similarity Score
    Book E: 0.00 Similarity Score
```

We can interpret the results as follows:

➤ **Book B (0.82):** This is a very strong recommendation. Why? Because Users 0 and 1 both bought Book A and Book B. The algorithm detected this strong pattern of co-purchase. If you're viewing Book A, Book B is the most logical next purchase.

➤ **Book C (0.41):** This is a moderate recommendation. While User 1 bought both, User 0 did not, and User 2 bought C but not A. The signal is mixed, so it gets a lower score.

➤ **Books D & E (0.00):** These have a score of 0. This is a crucial finding. It means there is no overlap between the customers who bought A and the customers who bought D or E. They appeal to completely different segments. The algorithm correctly identifies that recommending Book D to a Book A buyer would likely be a waste of screen space.

This simple geometric concept, measuring the angle between vectors, is the foundation of the recommendation engines used by Netflix, Amazon, and Spotify. By treating customer behavior as a vector, they can mathematically calculate taste and personalize the experience for millions of users instantly.

MAXIMIZING YIELD WITH CONSTRAINED OPTIMIZATION

Investment isn't just about picking winners; it's about following the rules. Banks, insurance companies, and pension funds operate under strict regulatory constraints. They want to maximize profit, but they must maintain a certain level of safety. It's a delicate balancing act: too much caution and you miss out on growth; too much risk and you violate the rules.

For this scenario, let's consider the business problem where you are a portfolio manager at a regional bank with $10 million to lend. You have three "buckets" you can deploy this capital into, each with a different risk-return profile:

➤ **Home mortgages:** These are very safe but offer a modest 3% return.

➤ **Small business loans:** These are riskier but offer a high 7% return.

➤ **Corporate bonds:** These sit in the middle, with medium risk and a 5% return.

Your goal is simple: maximize the total return on that $10 million. However, you are bound by strict internal risk policies that force diversification:

➤ **Constraint 1 (liquidity):** At least 50% of the total capital must be safe in home mortgages.

➤ **Constraint 2 (risk cap):** No more than 20% of the total capital can be exposed to high-risk small business loans.

➤ **Constraint 3 (diversity):** At least 10% must be allocated to corporate bonds to ensure market exposure.

This is a textbook linear programming problem like you have already worked on in the previous chapters. You need to maximize your total yield. If M is the portion in mortgages, S is the portion in small business loans, and C is the portion in corporate bonds, the return is as follows:

```
Return = 0.03*M + 0.07*S + 0.05*C
```

Subject to:

➤ M >= 0.5 (liquidity rule)

➤ S <= 0.2 (risk cap rule)

➤ C >= 0.1 (diversity rule)

➤ M + S + C = 1 (The budget constraint: we must invest exactly 100% of our capital)

We will use `scipy.optimize.linprog` to find the exact optimal mix. This tool will look at the different yields, push right up against the constraints where it makes sense, and tell us precisely how many dollars to put in each bucket to squeeze out every possible cent of profit while staying fully compliant. Listing 8-3 builds out this optimization problem according to the constraints.

LISTING 8-3: OPTIMIZING A CONSTRAINED PORTFOLIO

```python
import numpy as np
from scipy.optimize import linprog
import matplotlib.pyplot as plt

# Total Capital
total_capital = 10_000_000

# 1. Define the Objective Function (Maximize Return)
# Returns: Mortgages (3%), Small Biz (7%), Corp Bonds (5%)
```

```
# Since linprog minimizes, we use negative returns
c = [-0.03, -0.07, -0.05]

# 2. Define Constraints
# Variables: x0=Mortgages, x1=Small Biz, x2=Corp Bonds (as percentages of total)

# Inequality Constraints (A_ub * x <= b_ub)
# Constraint 1: Mortgages >= 50%  ->  -x0 <= -0.5
# Constraint 2: Small Biz <= 20%  ->   x1 <=  0.2
# Constraint 3: Corp Bonds >= 10% ->  -x2 <= -0.1
A_ub = [
    [-1,  0,  0], # Mortgages >= 0.5
    [ 0,  1,  0], # Small Biz <= 0.2
    [ 0,  0, -1]  # Corp Bonds >= 0.1
]
b_ub = [-0.5, 0.2, -0.1]

# Equality Constraint (A_eq * x = b_eq)
# All weights must sum to exactly 1 (100% of capital)
A_eq = [[1, 1, 1]]
b_eq = [1]

# Bounds: Each weight must be between 0 and 1
bounds = [(0, 1), (0, 1), (0, 1)]

# 3. Run the Optimizer
result = linprog(c, A_ub=A_ub, b_ub=b_ub, A_eq=A_eq, b_eq=b_eq, bounds=bounds,
method='highs')

# 4. Interpret Results
if result.success:
    weights = result.x
    optimal_return = -result.fun # Convert back to positive

    allocations = weights * total_capital

    print("--- Optimal Portfolio Allocation ---")
    print(f"Home Mortgages:      ${allocations[0]:,.0f} ({weights[0]:.0%})")
    print(f"Small Business Loans: ${allocations[1]:,.0f} ({weights[1]:.0%})")
    print(f"Corporate Bonds:      ${allocations[2]:,.0f} ({weights[2]:.0%})")
    print("---")
    print(f"Total Projected Return: ${optimal_return * total_capital:,.0f}
({optimal_return:.1%})")

    # --- Visualization ---
    labels = ['Mortgages (3%)', 'Small Biz (7%)', 'Corp Bonds (5%)']
    colors = ['lightblue', 'salmon', 'lightgreen']

    plt.figure(figsize=(7, 7))
    plt.pie(weights, labels=labels, autopct='%1.1f%%', colors=colors, startan-
gle=140)
    plt.title('Optimal Capital Allocation')
    plt.show()
else:
    print("Optimization failed.")
```

To maximize return using `linprog` (which is designed to minimize), we use the negative trick: we set our objective function coefficients (c) to the negative of our expected returns (e.g., `-0.07` instead of `0.07`). Minimizing negative profit is mathematically identical to maximizing positive profit.

The code in Listing 8-3 requires careful formatting to properly lay out the constraints. `linprog` expects all inequality constraints to be in the less than or equal to (`<=`) form. For constraints, such as mortgages must be at least 50%, we flip the inequality by multiplying both sides by –1 (so x `>=` 0.5 becomes `-x <= -0.5`). We pack these rules into the `A_ub` matrix (the coefficients) and `b_ub` vector (the limits). Finally, we define an equality constraint (`A_eq`, `b_eq`) so that the sum of our investments equals exactly 100% of our capital. Here is the result of running this code:

```
--- Optimal Portfolio Allocation ---
Home Mortgages:      $5,000,000 (50%)
Small Business Loans: $2,000,000 (20%)
Corporate Bonds:     $3,000,000 (30%)
---
Total Projected Return: $440,000 (4.4%)
```

The results reveal the logical strategy the algorithm discovered. You could imagine first, the algorithm allocated to the highest-yielding asset, small business loans (7%). It wanted to put 100% of the money there, but the risk cap constraint stopped it at 20%. So, it maxed out that bucket with $2 million.

Next, it had to satisfy the liquidity rule: at least 50% must be in home mortgages (3%). Even though this is the lowest-yielding asset, the constraint forces the algorithm to allocate $5 million here.

That left 30% of the capital ($3 million) unused. The algorithm looked at its remaining options. Corporate bonds (5%) pay better than mortgages, so it poured the entire remaining balance into bonds. It didn't put a single extra dollar into mortgages beyond the bare minimum required by law.

The final result is a portfolio that yields 4.4%. This is the mathematical ceiling. No other combination of investments can legally yield a higher return. This gives the portfolio manager confidence: they aren't guessing at the allocation; they have mathematically proven the best possible strategy given their constraints. Figure 8-2 shows the final allocation as a pie chart.

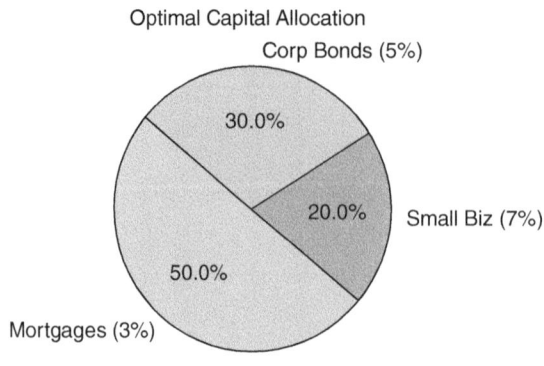

FIGURE 8-2: Loan allocation.

This solution illustrates the true power of linear programming in finance. It moves beyond simple intuition by correctly identifying the high-yield small business loans as attractive and further handles the complex balancing act required by the remaining capital. The resulting 4.4% return isn't just an estimate; it is the mathematical ceiling. No other allocation exists that satisfies your risk policies while generating a higher return, giving you the certainty needed to defend the strategy to the board.

QUALITY CONTROL WITH HYPOTHESIS TESTING

Quality control is the unsung hero of operations. Whether you are manufacturing steel bolts, bottling soda, or processing insurance claims, consistency is the gold standard. A slight deviation in a manufacturing process can compound into millions of dollars in wasted product, recalls, or reputational damage.

Imagine you manage a factory that produces high-precision steel bolts. Each bolt is designed to have a diameter of 10 mm with a minimal tolerance to fit perfectly into an engine assembly. You have just installed a new, expensive manufacturing machine. To validate it, you run a pilot batch of 50 bolts. You measure them with digital calipers and calculate the average: 10.02 mm.

This leaves you with a critical dilemma. Is that 0.02 mm difference just random noise (meaning the machine is fine)? Or is the machine actually calibrated incorrectly (meaning you are about to manufacture thousands of defective bolts)? If you shut down the line for recalibration when it wasn't needed, you burn money on downtime. If you don't shut it down when it is needed, you ship defective product. You need a mathematical way to distinguish signal from noise.

First, let's define what we hope to test:

➤ **Null hypothesis:** The machine is calibrated correctly. The true population mean is still 10 mm. The 10.02 mm deviation in the sample is just random noise.

➤ **Alternative hypothesis:** The machine is calibrated incorrectly. The true population mean is not 10 mm.

You can use the tool of hypothesis testing, t-testing, and `scipy.stats` to build out this decision-making process. Listing 8-4 shows this example.

LISTING 8-4: AUTOMATING QUALITY CONTROL CHECKS

```
import numpy as np
import matplotlib.pyplot as plt
from scipy import stats

# 1. Generate Mock Data from the "New Machine"
np.random.seed(42)
target_diameter = 10.0

# We simulate 50 bolts.
# The "Truth" (unknown to the manager) is that the machine IS slightly off
(mean=10.02).
# Standard deviation is 0.05mm.
sample_bolts = np.random.normal(loc=10.02, scale=0.05, size=50)
```

```
# 2. Analyze the Sample
sample_mean = np.mean(sample_bolts)
print(f"Target Diameter: {target_diameter}mm")
print(f"Sample Mean:     {sample_mean:.3f}mm")

# 3. Run the Hypothesis Test (One-Sample T-Test)
# We compare our sample array against the known population mean (10.0)
t_stat, p_value = stats.ttest_1samp(sample_bolts, popmean=target_diameter)

print(f"P-Value: {p_value:.4f}")

# 4. Make the Decision
if p_value < 0.05:
    print("DECISION: STOP THE LINE. The deviation is statistically significant.")
else:
    print("DECISION: CONTINUE. The deviation is likely random noise.")

# --- Visualization ---
plt.figure(figsize=(10, 6))
plt.hist(sample_bolts, bins=15, color='gray', alpha=0.7, label='Sample
Distribution')
plt.axvline(target_diameter, color='green', linewidth=2, linestyle='--',
label='Target (10mm)')
plt.axvline(sample_mean, color='red', linewidth=2, label=f'Sample Mean ({sample_
mean:.3f}mm)')
plt.title('Quality Control: Target vs. Actual Performance')
plt.xlabel('Bolt Diameter (mm)')
plt.legend()
plt.grid(True, alpha=0.3)
plt.show()
```

In Step 1, we simulate the actual data. We create 50 bolts using `np.random.normal`, giving them a true mean of `10.02`. This allows us to test if our statistical tool is sensitive enough to catch this tiny error. If you were implementing this example in our own manufacturing, you would provide a list of all of your measurements here.

In Step 2, we review the generated sample by looking at the mean of our simulated data.

Step 3 is the core of the solution. We use `stats.ttest_1samp`. This function takes our array of measurements and the value we expect them to match (`10.0`). It calculates how many standard errors away our sample mean falls from the target.

Finally, Step 4 automates the business logic. Instead of leaving the interpretation up to a tired shift manager, we hard-code the significance level (`0.05`). If the p-value drops below this threshold, the script explicitly prints STOP THE LINE, removing ambiguity from the decision.

The output from this code is as follows:

```
Target Diameter: 10.0mm
Sample Mean:     10.013mm
P-Value: 0.0548
DECISION: CONTINUE. The deviation is likely random noise.
```

The results of this simulation are fascinatingly tricky. Even though we (the programmers) know the machine is slightly off (set to 10.02), the p-value came back at 0.0548.

This is just barely above our 0.05 threshold. What does this mean? It means that getting a sample average of 10.013 mm when the machine is perfect would happen about 5.5% of the time just by chance. Because this probability is higher than our 5% risk tolerance, we fail to reject the null hypothesis. The script tells us to continue production. You can see this in visual form in Figure 8-3.

The sold line in Figure 8-3 is visibly separated from the dashed target line, but the statistics suggest it might still be noise. Again, you will notice this looks like the normal distribution, and that's by no accident. While in this example we generated the data to look like the normal distribution, we are also relying on the normal distribution to implement the t-test.

This example teaches a valuable lesson in analytics: statistical significance is not the same as practical truth. With a sample size of only 50, our test wasn't quite sensitive enough to confidently flag the tiny 0.02 mm error. If the quality manager wanted to catch such small deviations, they would need to increase the sample size (perhaps to 100 bolts) to give the test more power.

PREDICTING EMPLOYEE ATTRITION WITH LOGISTIC REGRESSION

Employee turnover is often called the "silent killer" of business growth. While a lost sale hurts immediately, a lost employee hurts indefinitely. Replacing a skilled knowledge worker can cost up to 200% of their annual salary when you factor in recruiting fees, onboarding time, and the "brain drain" of institutional knowledge walking out the door.

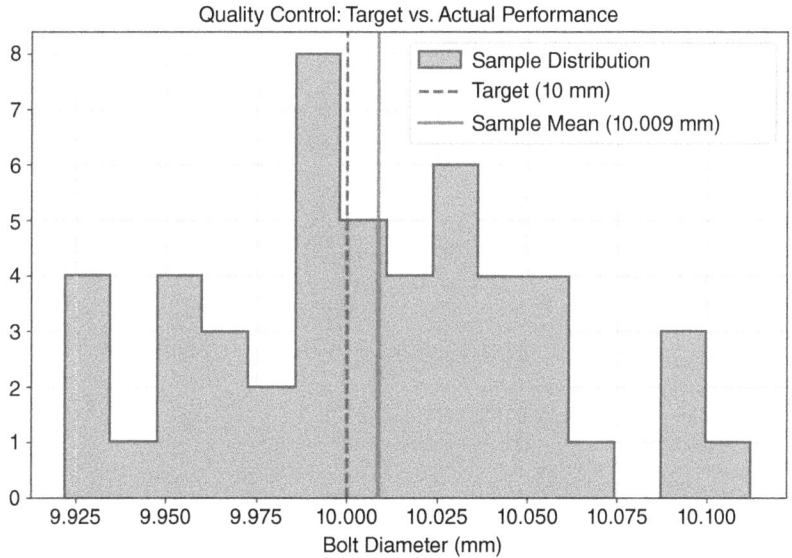

FIGURE 8-3: The distribution of our sample bolts.

Most HR departments operate reactively. They conduct exit interviews to find out why someone left, but by then, it's too late. HR directors often have a "gut feeling" about who might be at risk— perhaps the person with the grueling 90-minute commute or the mid-level manager who hasn't been promoted in three years. But gut feelings don't scale, and they don't justify the budget for retention programs. To intervene proactively, you need a mathematical early warning system.

In this business problem, you are the HR Director at a growing tech firm. You have historical data on 500 employees, including their tenure (years at the company), Last_Performance_Score (1–10), and Commute_Distance (miles). You need to answer two specific questions for the executive team:

➤ **The why:** Exactly how much does a long commute increase the risk of an employee quitting? Is it a minor nuisance or a major driver?

➤ **The who:** Can you flag specific high-risk employees right now, before they hand in their resignation, so you can offer them remote work options or a raise?

This is a classification problem. Unlike our previous examples where we predicted a continuous number (like profit or sales), here we are predicting a category: Quit vs. Stay. We will use logistic regression. Unlike linear regression, which tries to fit a straight line through the data, logistic regression fits an S-curve (sigmoid function). This is crucial because it constrains the output to be between 0 and 1, giving us a clean probability of turnover.

We will use statsmodels.Logit to build the model. We'll start by simulating a realistic dataset where we know the "ground truth" rules of attrition. Then, we'll train our model to see if it can rediscover those rules. Finally, we will interpret the model's coefficients by converting them into odds ratios, a metric that translates complex log-odds into simple business logic. Listing 8-5 implements this solution.

LISTING 8-5: MODELING EMPLOYEE ATTRITION RISK

```
import numpy as np
import statsmodels.api as sm

# 1. Generate Synthetic HR Data (500 Employees)
np.random.seed(101)
num_employees = 500

# Independent Variables
# Randomly generate realistic employee data
tenure = np.random.uniform(1, 10, num_employees)
perf_score = np.random.uniform(1, 10, num_employees)
commute_dist = np.random.uniform(1, 50, num_employees) # 1 to 50 miles

# Define "True" Risk Logic (Hidden Reality)
# We simulate a world where long commutes increase risk,
# while high tenure and high performance decrease it.
# (These coefficients create the "log-odds" of quitting)
log_odds = -1.5 + (0.08 * commute_dist) - (0.3 * tenure) - (0.1 * perf_score)

# Convert log-odds to probability using the Sigmoid function
prob_quit = 1 / (1 + np.exp(-log_odds))
```

```python
# Generate Outcomes (1=Quit, 0=Stay) based on that probability
quit_status = (np.random.rand(num_employees) < prob_quit).astype(int)

# 2. Fit Logistic Regression Model
y = quit_status
# Create our X matrix with all three variables
X = np.column_stack((commute_dist, tenure, perf_score))
X = sm.add_constant(X) # Add intercept

# We use Logit instead of OLS for binary (Yes/No) outcomes
model = sm.Logit(y, X).fit(disp=0) # disp=0 suppresses the training log output

# 3. Interpret Coefficients (Odds Ratios)
# We exponentiate the coefficients to get readable Odds Ratios
params = model.params
odds_ratios = np.exp(params)

print("\n--- Risk Factor Analysis (Odds Ratios) ---")
print(f"Baseline (Intercept): {odds_ratios[0]:.4f}")
print(f"Commute Distance:      {odds_ratios[1]:.4f} (Impact of +1 mile)")
print(f"Tenure (Years):        {odds_ratios[2]:.4f} (Impact of +1 year)")
print(f"Performance Score:     {odds_ratios[3]:.4f} (Impact of +1 point)")

# 4. Predict Risk for Specific Employees
# Employee A: 5 mile commute, 8 years tenure, 9 performance (Loyal profile)
# Employee B: 45 mile commute, 2 years tenure, 5 performance (Risk profile)
new_employees = np.array([
    [1, 5, 8, 9],
    [1, 45, 2, 5]
])
predicted_probs = model.predict(new_employees)

print("\n--- Attrition Risk Predictions ---")
print(f"Employee A (Happy):   {predicted_probs[0]:.1%} probability of quitting")
print(f"Employee B (At Risk): {predicted_probs[1]:.1%} probability of quitting")
```

Listing 8-5 has several steps. In Step 1, we generate a realistic, synthetic dataset to simulate our customer base. After setting a `np.random.seed(101)` to ensure our results are reproducible, we create data for 500 employees. We define our independent variables (*X* variables) first: `tenure` is a uniform continuous value between 1 and 10, `perf_score` is between 1 and 10, and `commute_dist` is between 1 and 50 miles. The most complex part of this step is generating our dependent *y* variable, `quit_status`. To do this, we invent a "true" (but hidden) relationship using `log-odds`. We create a `log_odds` variable for each employee based on our secret formula (where `commute_dist` increases risk and tenure decreases it). We convert these log-odds into a clean probability between 0 and 1 using the sigmoid function, `1 / (1 + np.exp(-log_odds))`. Finally, we simulate the binary outcome for each employee with a "coin flip" using `np.random.rand() < prob_quit`.

In Step 2, we prepare this data for the `statsmodels` library. We assign our binary `quit_status` array to *y*. We then use `np.column_stack` to bundle our independent variables into a single *X* matrix. The line `X = sm.add_constant(X)` is a critical step that adds the intercept column, allowing the model to calculate a baseline probability. We fit the model using `sm.Logit(y, X)`, which tells `statsmodels` to fit a logistic curve.

Finally, in Steps 3 and 4, we interpret and apply the model. We convert the raw coefficients into odds ratios by exponentiating them (`np.exp(params)`), which makes them interpretable for business decisions. We then use `model.predict()` to feed new, hypothetical employee data into our trained model to see their calculated risk scores.

The output provides two levels of insight—strategic drivers and individual risk scores:

```
--- Risk Factor Analysis (Odds Ratios) ---
Baseline (Intercept): 0.2848
Commute Distance:     1.0851 (Impact of +1 mile)
Tenure (Years):       0.6981 (Impact of +1 year)
Performance Score:    0.9050 (Impact of +1 point)

--- Attrition Risk Predictions ---
Employee A (Happy):   1.0% probability of quitting
Employee B (At Risk): 76.9% probability of quitting
```

First, let's look at the odds ratios, which tell us the power of each factor. The commute distance (1.085) is the most critical finding for the HR Director. It translates to: For every single extra mile of commute, an employee's odds of quitting increase by 8.5% (since 1.085 is 8.5% higher than 1.0). While one mile seems small, this effect compounds. A 30-mile commute creates a massive accumulation of risk compared to a 5-mile commute. Tenure (0.698), on the other hand, is below 1.0, meaning it has a protective effect. It tells us: For every additional year an employee stays, their odds of quitting decrease by about 30% (1.0—0.698). Loyalty begets loyalty; the longer someone stays, the stickier they become. Finally, the performance score (0.910) shows that high performers are also slightly less likely to leave, with each point of performance reducing quit odds by about 9%.

Second, we look at the predictions, where the model allows us to score every single person in the company. Employee A represents our safe profile: they live close (five miles), have been here a long time (eight years), and perform well. The model assigns them a tiny 1.0% probability of leaving, so we don't need to worry about them right now. Employee B, however, is a flashing red light. They have a long commute (45 miles) and have only been with the company for two years. The model calculates their quit probability at 76.9%. Even though they haven't said anything, the math suggests they are already halfway out the door.

Let's see what this would look like if we were to create a visualization. By combining the code from Listings 8-5 and 8-6, you can see how employees tenure might be affected.

LISTING 8-6: VISUALIZING EMPLOYEE ATTRITION RISK

```
import numpy as np
import matplotlib.pyplot as plt
import statsmodels.api as sm

# Run the code from Listing 8-5 first!
```

```
# --- Visualization: The Risk Curve ---
plt.figure(figsize=(10, 6))

# 1. Create a range of Commute Distances to plot (0 to 60 miles)
commute_range = np.linspace(0, 60, 100)

# 2. Define "Average" values for other variables to isolate the effect of Commute
avg_tenure = np.mean(tenure)
avg_perf = np.mean(perf_score)

# 3. Predict probability for each commute distance
# We manually build the matrix to ensure it has 4 columns: [Const, Commute,
Tenure, Perf]
# We do this because sm.add_constant() can get confused when other columns
are constant.
X_plot = np.column_stack((
    np.ones(100),              # The Intercept (Must be first)
    commute_range,            # Variable 1: Commute (Changing)
    np.full(100, avg_tenure), # Variable 2: Tenure (Fixed)
    np.full(100, avg_perf)    # Variable 3: Performance (Fixed)
))

# Get probabilities from the model
probs = model.predict(X_plot)

# 4. Plot the Curve
plt.plot(commute_range, probs, color='darkred', linewidth=3, label='Attrition
Risk Curve')

# 5. Plot Specific Employees
# We also need to ensure single predictions have the intercept added manually
emp_a_data = [1, 5, 8, 9]
emp_b_data = [1, 45, 2, 5]

emp_a_prob = model.predict(emp_a_data)[0]
emp_b_prob = model.predict(emp_b_data)[0]

plt.scatter([5], [emp_a_prob], color='green', s=100, zorder=5, label='Employee A
(Safe)')
plt.scatter([45], [emp_b_prob], color='red', s=100, zorder=5, label='Employee B
(Flight Risk)')

plt.annotate(f"{emp_a_prob:.0%}", (5, emp_a_prob), xytext=(4, emp_a_prob+0.05),
fontsize=16)
plt.annotate(f"{emp_b_prob:.0%}", (45, emp_b_prob), xytext=(44, emp_b_prob-0.06),
fontsize=16)

plt.title('Impact of Commute on Employee Flight Risk')
plt.xlabel('Commute Distance (Miles)')
plt.ylabel('Probability of Quitting')
plt.axhline(0.5, color='gray', linestyle='--', alpha=0.5, label='50% Risk
Threshold')
plt.legend()
plt.grid(True, alpha=0.3)
plt.show()
```

The visualization code in Listing 8-6 is designed to transform our abstract logistic regression model into a tangible, strategic tool. Our goal is to visualize exactly how commute distance impacts the probability of attrition, while holding all other factors constant.

To achieve this, we first create a hypothetical scenario. In Steps 1 and 2, we define a range of possible commute distances from 0 to 60 miles using `np.linspace`. To isolate the specific effect of the commute, we calculate the average tenure and average performance score from our dataset and "freeze" them at these values. This allows us to answer the question: For an average employee, how does adding miles to their commute change their risk?

In Step 3, we prepare the input matrix `X_plot` for the model. This requires careful manual construction using `np.column_stack`. The model expects four specific inputs for every prediction: the intercept (a column of ones), the variable commute distances, and the fixed columns for tenure and performance. By building this matrix explicitly, we ensure the dimensions align perfectly with the model's trained coefficients. We then pass this matrix to `model.predict()`, which returns the probability of quitting for every mile along our 60-mile range.

In Step 4, we plot this resulting data as a smooth, dark red line (`plt.plot`), creating the "risk curve." This curve allows us to visually identify the tipping point where the risk of quitting accelerates.

Finally, in Step 5, we overlay the specific cases of Employee A and Employee B to make the data personal. We manually define their data points remembering to include the leading 1 for the intercept and predict their individual probabilities. We use `plt.scatter` to plot them as distinct points on the graph: a dot for the safe Employee A and a dot for the at-risk Employee B. We finish the chart, as seen in Figure 8-4, by adding annotations to display their exact probabilities and a dashed gray line at the 50% mark (`plt.axhline`), providing a clear visual threshold for high-risk employees.

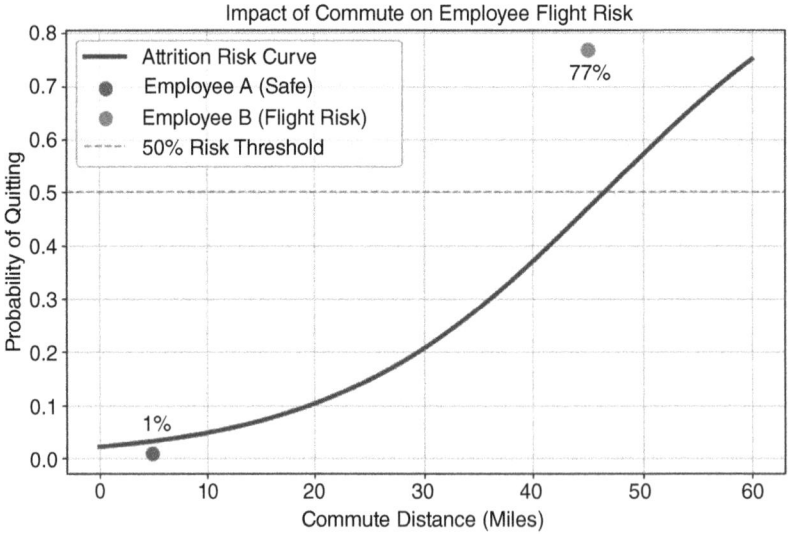

FIGURE 8-4: Visualization of the impact of commute on employee attrition.

This transforms HR from a reactive function into a strategic one. Instead of waiting for resignation letters, you can now generate a risk report every Monday morning. You can specifically target Employee B and others like them with retention offers, perhaps a remote workday to mitigate that commute risk, before it's too late.

SUMMARY

This chapter marked a significant transition in your journey. You have moved from stocking a toolbox with individual mathematical concepts, such as calculus, linear algebra, optimization, and statistics, to entering the workshop and building complete, functional solutions. You stepped away from abstract equations and tackled the messy, word problem reality of business, where the goal is not just to solve a problem, but to save time, reduce risk, and increase efficiency.

The chapter began by constructing a dynamic loan amortization engine. By combining financial math with Python's logic, you built a tool capable of simulating decades of payments in milliseconds, revealing how a small strategic change in monthly cash flow could save hundreds of thousands of dollars in interest. It then pivoted to marketing, using the geometric concept of vectors and cosine similarity to build a recommendation engine from scratch, translating raw transaction data into personalized customer experiences.

From there, you applied constrained optimization to finance, proving that the "best" portfolio isn't just a guess, but a mathematical certainty that balances yield against strict risk policies. The chapter then moved to the factory floor, using hypothesis testing to replace "gut feeling" with a rigorous, automated quality control system that distinguishes true manufacturing errors from random noise. Finally, you empowered HR with predictive analytics, using logistic regression to turn employee data into a proactive early warning system for attrition. You now possess a library of reusable models and, more importantly, the ability to look at a business challenge, identify the underlying math, and build the exact tool required to solve it.

CONTINUE YOUR LEARNING

This chapter was about bridging the gap between theory and practice. To truly master the art of applying mathematics to business, you need to deepen your understanding of both the modeling techniques (the how) and the strategic thinking behind them (the why).

Strategic Modeling and Business Analytics

➤ *The Signal and the Noise* by Nate Silver: An excellent, high-level read on the philosophy of prediction and why models fail. It provides crucial context for the "Quality Control" and "Forecasting" sections, emphasizing the difference between probabilistic thinking and absolute certainty.

➤ *Mastering 'Metrics: The Path from Cause to Effect* by Joshua Angrist and Jörn-Steffen Pischke: An accessible, amazing guide to the econometric tools used in business (random assignment, regression, instrumental variables, regression discontinuity, and differences-in-differences). An essential text for understanding causal inference.

➤ *How Not to Be Wrong: The Power of Mathematical Thinking* by Jordan Ellenberg: This book is a great for applying simple math to real-world problems.

Technical Documentation

➤ `Scipy.optimize:` The official documentation for the optimization library we used. It covers not just linprog (linear programming) but also nonlinear solvers that are essential for more complex supply chain and logistics problems. `https://docs.scipy.org/doc/scipy/reference/optimize.html`

➤ `Statsmodels:` While Scikit-Learn is great for machine learning, `Statsmodels` is the gold standard for the type of statistical inference used in the "Employee Attrition" section. Their documentation on generalized linear models (GLMs) is the next step in your learning journey. `https://www.statsmodels.org/stable/index.html`

PART 3
Visualizing the Numbers

➤**Chapter 9:** Illustrating Time-series and Linear Data

➤**Chapter 10:** Illustrating Cross-sectional Data

➤**Chapter 11:** Illustrating Alternative Data Types

Illustrating Time-series and Linear Data

A spreadsheet with 10,000 rows is impenetrable. No matter how rigorous your math or how optimized your code, if you hand a stakeholder a wall of numbers, your insight is lost. A single, well-crafted chart, however, can lead to a boardroom decision in seconds.

Visualization isn't just about making things pretty; it's about making complex patterns instantly recognizable. It is the final, critical mile of the analytics marathon. Visualization is where you translate your digital findings into human understanding.

This chapter moves beyond simply calculating numbers to telling visual stories. You will master the skill of matching the right chart to the right data structure. You'll learn to reveal hidden correlations using scatterplots, highlight clear trends with regression lines, smooth out noisy time-series data to see the signal in the noise, and effectively compare disparate metrics on a single view. The primary toolkit for this is Matplotlib, Python's foundational plotting library, along with the powerful visualization capabilities built directly into Pandas.

UNDERSTANDING YOUR DATA STRUCTURE

Before you write a single line of plotting code, you must ask yourself one simple but critical question: What is the shape of my data? The structure of your dataset dictates the visualization you should use. Misunderstanding this structure is the most common reason charts fail to communicate effectively. In business analytics, data almost always falls into one of three fundamental categories: cross-sectional, time-series, or panel data.

Cross-sectional data is a snapshot in time. Imagine taking a single photo of your business right now. You aren't looking at history; you are looking at the relationship between many different subjects at this specific moment. A classic example is analyzing the total revenue for all 50 states in 2023. You aren't asking how the revenue changed (that's time-series)—you are instead asking how California compares to Texas right now.

Other common examples of cross-sectional data include comparing the profitability of different product lines, analyzing customer demographics from a single survey, or looking at the inventory levels across all your warehouses today. For this type of data, your best visualization tools are bar charts for direct comparison, scatterplots or line plots (like Figures 9-1 and 9-2) for finding relationships between variables (e.g., do states with higher populations generate more revenue?), and histograms for understanding distributions (like the one created in Figure 9-3).

Time-series data is a history. It tracks a single subject over multiple time intervals. Instead of comparing California to Texas, you are looking at California's revenue in January, February, March, and so on. The goal here is to identify trends, seasonality, and cycles. A daily stock price chart is the quintessential time-series example. Other business examples include your company's monthly recurring revenue (MRR) over the last five years, the daily foot traffic in a flagship store, or the hourly server load on your website. Because the sequence of data points matters, the line chart is the undisputed king of time-series visualization. It visually connects the dots to show the flow of time. For example, Figure 9-1 shows the Consumer Price Index (CPI) overtime. Area charts are also powerful here, especially for showing cumulative totals.

Panel data is the combination of cross-sectional and time-series data. It tracks multiple subjects over multiple time periods. Imagine a spreadsheet where you have daily stock prices (time-series) for Apple, Google, and Microsoft (cross-sectional). This is the richest and most complex form of data because it allows you to answer multi-dimensional questions such as: Which tech giant has grown the fastest over the last decade?

Visualizing panel data requires care to avoid clutter. Multiple line charts on a single axis work well if you have only a few subjects (e.g., three to five companies). If you have data for all 50 states over 10 years, a spaghetti chart of 50 lines will be unreadable. In that case, faceted plots (or small multiples), where each subject gets its own small chart, are often the best choice. In Python, the quickest way to identify your data structure is to inspect the index of your Pandas DataFrame. If the index consists of dates or times (e.g., January 1, 2023), you are almost certainly dealing with time-series data. If it consists of IDs, names, or categories (e.g., product A, user 101), it is likely cross-sectional. Recognizing this distinction is the first step to building a chart that tells the truth.

Cross-sectional Data

Let's put this into practice. Before we start plotting, we need to load our data and understand its shape, which is done in Listing 9-1. We'll use a dataset of vegetable prices, which captures a snapshot of retail prices for various produce items in 2022. Since this data represents a single point in time across many different items, it is a classic example of cross-sectional data.

> **Note** Note that Listing 9-1 uses a CSV file called `Vegetable-Prices-2022.csv` found on GitHub. This file along with the files used in the other listings in this chapter are included in the downloadable files for this book, located on the Wiley site at `https://github.com/bkrayfield/Applied-Math-With-Python`.

LISTING 9-1: INSPECTING DATA STRUCTURE WITH PANDAS

```
import pandas as pd

# Load the dataset directly from the source
url = "https://github.com/bkrayfield/Applied-Math-With-Python/raw/refs/heads/main/
Data/Vegetable-Prices-2022.csv"

veg_prices = pd.read_csv(url)

# Inspect the first few rows to understand the structure
print(veg_prices.head())

# Check the index and columns
print("\n--- Data Info ---")
print(veg_prices.info())
```

In Listing 9-1, we begin by importing the pandas library. We define the source URL for our dataset and use the pd.read_csv() function to load it into a DataFrame named veg_prices. To understand what we just loaded, we use two key methods: head() prints the first five rows, giving us a visual snapshot of the columns and values, while info() provides a structural summary, detailing the data types (integers, floats, objects) and counting non-null values to check for missing data.

Running this code gives us a first look at the data:

```
           Vegetable    Form  RetailPrice RetailPriceUnit   Yield  CupEquivalentSize
CupEquivalentUnit  CupEquivalentPrice
0     Acorn squash    Fresh       1.2136    per pound  0.4586             0.4519
pounds               1.1961
1        Artichoke    Fresh       2.4703    per pound  0.3750             0.3858
pounds               2.5415
2        Artichoke   Canned       3.4498    per pound  0.6500             0.3858
pounds               2.0476
3        Asparagus    Fresh       2.9531    per pound  0.4938             0.3968
pounds               2.3731
4        Asparagus   Canned       3.4328    per pound  0.6500             0.3968
pounds               2.0958
```

By looking at the .head() output, we can confirm our data type:

➤ **There are no dates:** The rows are indexed by simple integers (0, 1, 2, ...), and there is no date column.

➤ **Categorical subjects:** Each row represents a distinct subject (a specific vegetable form, like fresh artichoke vs. canned artichoke).

➤ **Snapshot values:** The RetailPrice and Yield columns represent values at a fixed point in time (2022).

This confirms we are working with cross-sectional data. Knowing this immediately tells us which visualization tools to reach for: bar charts to compare prices between vegetables, or scatterplots to explore the relationship between the RetailPrice and Yield columns. You wouldn't use a line chart here, because there is no time progression to connect the dots.

Time-series Data

Now, let's look at a completely different structure. Listing 9-2 loads a dataset representing the Consumer Price Index (CPI). The CPI is the definitive yardstick for the cost of living in the United States. It measures the weighted average change in prices paid by urban consumers for a specific basket of goods and services, ranging from groceries and gasoline to housing and healthcare. When this index rises, it indicates that purchasing power is declining, a phenomenon we collectively call *inflation*. This data tracks a single metric (CPI) over many decades.

LISTING 9-2: INSPECTING TIME-SERIES DATA

```python
import pandas as pd
# Load the CPI dataset
cpi_url = "https://github.com/bkrayfield/Applied-Math-With-Python/raw/refs/heads/
main/Data/CPIAUCSL.csv"

# We tell pandas to parse the 'DATE' column as dates, not strings
cpi_data = pd.read_csv(cpi_url, parse_dates=['observation_date'])

# Set the Date as the index (Best Practice for Time Series)
cpi_data = cpi_data.set_index('observation_date')

print(cpi_data.head())
print("\n--- Data Info ---")
print(cpi_data.info())
```

In Listing 9-2, we load the dataset with `pd.read_csv`, but we add the `parse_dates=['observation_date']` argument. This instructs Pandas to convert the text strings in the DATE column into actual datetime objects during the load process. Immediately after, we call `set_index('observation_date')` to move this column to the index. This step is crucial for time-series data; it transforms the DataFrame from a simple list of rows into a time-aware structure, enabling powerful date-based plotting and resampling later. We verify this transformation by printing the head and info:

```
              CPIAUCSL
DATE
1947-01-01       21.48
1947-02-01       21.62
1947-03-01       22.00
1947-04-01       22.00
1947-05-01       21.95

--- Data Info ---
<class 'pandas.core.frame.DataFrame'>
DatetimeIndex: 925 entries, 1947-01-01 to 2024-01-01
Data columns (total 1 columns):
 #   Column    Non-Null Count  Dtype
---  ------    --------------  -----
 0   CPIAUCSL  925 non-null    float64
```

Looking at the output, the structure is distinct:

➤ **Datetime index:** The rows are no longer simple integers (0, 1, 2). The index is now a DatetimeIndex, meaning Pandas understands the sequence of time.

➤ **Single subject:** There is only one data column (CPIAUCSL), representing the value of the CPI.

➤ **Sequential history:** The data moves forward in monthly steps from 1947.

This is pure time-series data. Because we have a DatetimeIndex, we know that plotting cpi_data.plot() will automatically generate a line chart with a correctly formatted x-axis (years), showing the trend of inflation over time. A bar chart would be cluttered and inappropriate here.

Panel Data

Finally, let's look at the most complex data type. Listing 9-3 loads a dataset of energy prices. This data tracks the price of energy (in dollars per million Btu) across different sectors (residential, commercial, industrial) and fuel types over many years. It combines cross-sectional categories with time-series history.

LISTING 9-3: INSPECTING PANEL DATA

```
import pandas as pd
# Load the Energy Prices dataset
energy_url = "https://github.com/bkrayfield/Applied-Math-With-Python/
raw/refs/heads/main/Data/Energy_Prices,_Dollars_per_Million_Btu__
Beginning_1970_20251201.csv"
energy_data = pd.read_csv(energy_url)

energy_data = energy_data.melt(
    id_vars=['Year', 'GDP Deflator', 'Sector'],
    var_name='Fuel Type',
    value_name='Price'
)

print(energy_data[['Year','Sector','Fuel Type','Price']].head())
```

Because panel data is multidimensional, simply loading it isn't enough; we need to verify its layout. By printing the head(), we can observe the long format structure: the Sector and Fuel Type columns act as identifiers, while Year provides the time dimension. This confirms that multiple observations exist for each year, distinguishing this data from simple time-series data. The output should look like the following:

```
    Year          Sector Fuel Type  Price
0   1970      Commercial      Coal  $0.48
1   1970      Industrial      Coal  $0.53
2   1970     Residential      Coal  $1.43
3   1970  Transportation      Coal    NaN
4   1971      Commercial      Coal  $0.56
```

This is a classic example of panel data (specifically in "Long Format"):

➤ **Multiple entities:** We have a `Sector` column and a `Fuel Type` column. This is the cross-sectional part (comparing Residential vs. Industrial).

➤ **Time component:** We have a `Year` column. This is the time-series part.

➤ **Repeated observations:** We see Residential Natural Gas in 1970, and then again in 1971, 1972, and so on.

Recognizing this panel structure is crucial because a simple line chart won't work. If you plot Price vs. Year, Python will try to connect all the sectors into one jagged, messy line. Instead, this structure tells us we need to group or pivot the data. We would create a multiline chart, drawing one separate line for each sector, allowing us to compare their trends over time on a single set of axes.

You've now seen a little bit on different types of data. The rest of this chapter visualizes and works with time-series data. The next chapter then digs more into the cross-sectional data. This is followed with a dive into working with alternative types of data in Chapter 11.

VISUALIZING CHANGE OVER TIME (TIME-SERIES)

Let's start working more with time-series data. One of the most fundamental tasks in analytics is tracking how a metric evolves. Is revenue growing? Is inflation stabilizing? Are there seasonal patterns in our sales?

The line chart is the undisputed king of time-series visualization. By connecting data points with a continuous line, it emphasizes the flow of time and makes trends instantly visible.

Imagine you are an economist or a business analyst tracking inflation. You have decades of monthly Consumer Price Index (CPI) data (which we loaded in Listing 9-2 as `cpi_data`). The raw data can be jagged and noisy, making it hard to see the underlying trend. You want to visualize the long-term trajectory of inflation.

To solve this, you can use `matplotlib.pyplot` to plot the CPI data. Because we set the date as our index in Listing 9-2, plotting is incredibly straightforward. To add more value, Listing 9-4 also calculates and plots a rolling average (or moving average). This technique smooths out short-term fluctuations to reveal the clearer long-term trend.

LISTING 9-4: VISUALIZING TRENDS WITH SMOOTHING

```
import matplotlib.pyplot as plt

# 1. Prepare the Data
# We already loaded 'cpi_data' in Listing 9.2 with a DatetimeIndex
# Let's filter to a more recent period (e.g., from 2000 onward) for clarity
recent_cpi = cpi_data['2000':]

# 2. Calculate a Rolling Average
```

```
# We'll use a 12-month window to smooth out annual seasonality
rolling_avg = recent_cpi.rolling(window=12).mean()

# 3. Visualization
plt.figure(figsize=(12, 6))

# Plot the raw data as a faint, thin line
plt.plot(recent_cpi.index, recent_cpi['CPIAUCSL'],
        color='gray', alpha=0.4, linewidth=1, label='Monthly CPI (Raw)')

# Plot the smoothed trend as a bold, colored line
plt.plot(rolling_avg.index, rolling_avg['CPIAUCSL'], f
        color='blue', linewidth=2, label='12-Month Moving Average')

# Add formatting
plt.title('Consumer Price Index (CPI) Trend (2000-2024)')
plt.xlabel('Year')
plt.ylabel('CPI Value')
plt.legend()
plt.grid(True, linestyle='--', alpha=0.5)

plt.show()
```

In Listing 9-4, we visualize the Consumer Price Index (CPI), a classic time-series dataset that is often noisy due to month-to-month fluctuations. In Step 1, we filter the data to focus on the period from the year 2000 onward, making the recent trends easier to see. In Step 2, we apply a powerful technique called a *moving average*. By using .rolling(window=12).mean(), we tell Pandas to take a 12-month window of data, calculate the average, and then slide that window forward one month at a time. This smooths out the seasonal bumps, like holiday shopping spikes, revealing the true underlying direction of the economy.

The visualization code then layers these two perspectives onto a single chart. We plot the raw monthly data as a shaded line in the background. This preserves the detail without dominating the view. On top of that, we plot our 12-month moving average as a bold line. The resulting line plot can be seen in Figure 9-1.

The figure tells a clear story of signal versus noise. The light line (raw data) is erratic, jumping up and down every month. However, the solid line (the trend) cuts through the noise, showing a clear, smooth trajectory of inflation over the last two decades. This technique allows analysts to acknowledge the volatility of the moment while keeping stakeholders focused on the long-term direction.

Time-series Diagnostics

Before applying advanced models or attempting to forecast future trends, an analyst must perform a rigorous health check on their time-series data. Real-world data is rarely clean; it is often plagued by sensor failures, holiday gaps, and random outliers that can skew averages and break models. This section covers the essential diagnostic toolkit: cleaning data to ensure continuity, aggregating timelines to match business cycles, and summarizing distributions to understand volatility.

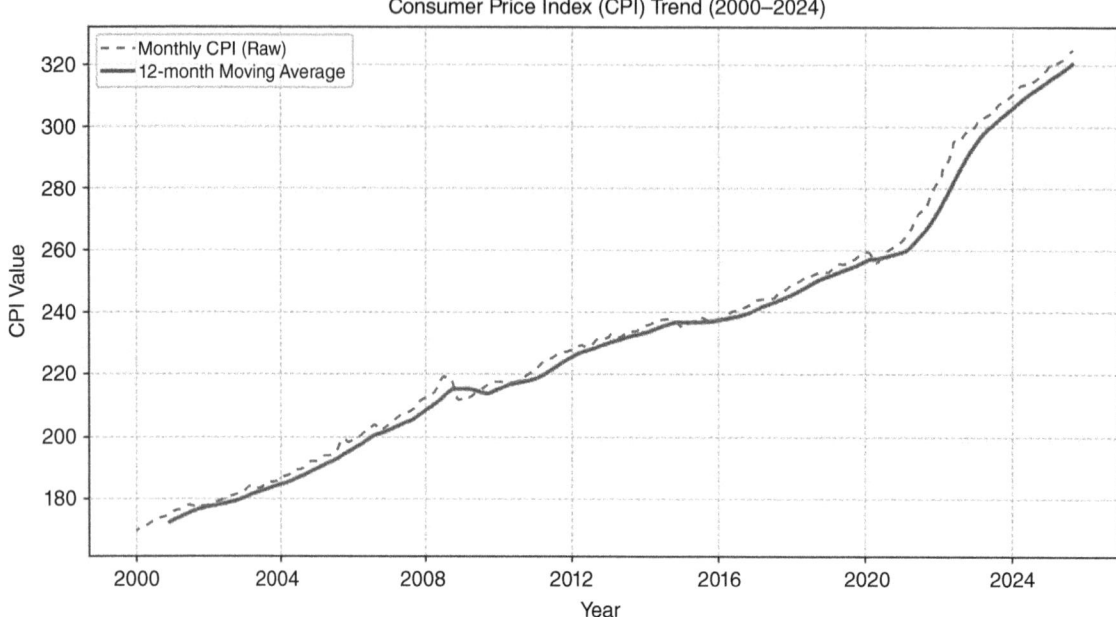

FIGURE 9-1: A line plot with a moving average.

Detecting Missing Values

Missing data in a time series poses a unique challenge because time is sequential. In a standard survey dataset, you might simply delete a row with missing values without much consequence. However, deleting a row in a time series breaks the continuity of the timeline, disrupting calculations like rolling averages or day-over-day changes. Instead of deletion, you must first identify these gaps.

Imagine you have daily sales data, but your point-of-sale system went down for a weekend, leaving NaN (not a number) gaps in your record. Listing 9-5 builds out this problem using synthetic data (data that is generated to simulate real data).

LISTING 9-5: FINDING AND FILLING MISSING TIME-SERIES DATA

```
import pandas as pd
import numpy as np
import matplotlib.pyplot as plt

# 1. Generate Synthetic Data with Gaps
np.random.seed(42)
dates = pd.date_range(start='2024-01-01', periods=20, freq='D')
# Create a base trend + some noise
values = np.linspace(100, 200, 20) + np.random.normal(0, 5, 20)
ts_data = pd.DataFrame({'Sales': values}, index=dates)

# Artificially create gaps to simulate a system failure
```

```
ts_data.iloc[5:8] = np.nan  # 3 days of missing data (indices 5, 6, 7)

# 2. Detect the Gaps
missing_count = ts_data['Sales'].isna().sum()
print(f"Missing Values Detected: {missing_count}")

# Show the rows with missing data
print("\n--- Rows with Missing Data ---")
print(ts_data[ts_data['Sales'].isna()])
```

Listing 9-5 first creates a synthetic dataset representing 20 days of sales. We then intentionally break the data by setting a three-day block to NaN (not a number). The key diagnostic tool here is .isna(), which returns a True or False value for every row. By chaining .sum(), we get a quick count of how many holes there are in our dataset. This is the first sanity check every analyst should run. The result of running the code in Listing 9-5 is as follows:

```
Missing Values Detected: 3

--- Rows with Missing Data ---
            Sales
2024-01-06    NaN
2024-01-07    NaN
2024-01-08    NaN
```

Now that we know what data is missing, we can talk about how to work with missing data.

Handling Missing Values

Once you know where the gaps are, you must fill them. One intuitive approach is to fill them with zeros, but for sales data, this is incorrect. In the business example, sales were likely not zero; we just failed to record them. A better approach is to estimate what those values would have been based on the trend. This is called *interpolation*.

Interpolation is more than just connecting the dots; it is a method of constructing new data points within the range of a discrete set of known data points. In business analytics, it assumes that the world doesn't change instantly and that values likely move smoothly from one point to another.

There are several ways to interpolate, and choosing the right one matters:

➤ **Linear interpolation (default):** This draws a straight line between two points. It's safe, simple, and usually good enough for short gaps like a weekend outage.

➤ **Time interpolation (method='time'):** This is a different version of linear interpolation that respects the distance between dates. If your data isn't evenly spaced (e.g., you have data for January 1, January 2, and January 10), time interpolation understands that January 10 is farther away and adjusts the slope accordingly. This is the gold standard for irregular time series.

➤ **Polynomial or spline interpolation:** These fit a curve (like a parabola) to the data rather than a straight line. While they can look smoother and more natural, they are risky. They can overshoot or undershoot, predicting wild spikes or drops in the gap that never actually happened.

Listing 9-5 creates synthetic times-series data. Because this data is dependent on time, we can create an example in Listing 9-6 that uses time interpolation.

LISTING 9-6: FILLING GAPS WITH TIME-BASED INTERPOLATION

```
# Run Listing 9-5 before running this code!
# 3. Handle Missing Values using Interpolation
# method='time' ensures the fill accounts for the actual time
distance between points
clean_data = ts_data.interpolate(method='time')

print(f"\nMissing Values After Cleaning: {clean_data['Sales'].isna().sum()}")

# --- Visualization ---
plt.figure(figsize=(10, 6))
# Plot the original data (with gaps) as dots
plt.plot(ts_data.index, ts_data['Sales'], 'o', label='Observed Data',
color='blue', markersize=8)
# Plot the cleaned data as a line to show how it bridges the gap
plt.plot(clean_data.index, clean_data['Sales'], '--', label='Interpolated Fill',
color='orange', alpha=0.7)

plt.title('Handling Missing Data with Time-Based Interpolation')
plt.xlabel('Date')
plt.ylabel('Sales')
plt.legend()
plt.grid(True, linestyle='--', alpha=0.5)
plt.show()
```

Listing 9-6 uses the `.interpolate()` function. Specifically, we use `method='time'`. This is critical for time-series data. It looks at the index (the dates) and draws a line between the valid points surrounding the gap. If the gap is three days long, it calculates the slope needed to get from the start to the end of that three-day period and fills in the missing days accordingly. This preserves the overall trend of the data without introducing artificial dips (like filling with 0) or plateaus (like filling with the mean). The result now is no missing data points. You can visualize what is happening in Figure 9-2.

The biggest risk with interpolation is hallucinating data. If you have a gap of three months in your sales data, drawing a straight line across it creates 90 days of fake data that implies a smooth, steady trend that likely never existed. As a rule of thumb, interpolate short gaps (to fix glitches) but treat long gaps as separate periods of analysis.

Aggregating and Resampling

Often, the raw data you collect is too granular for strategic decision-making. You might have data recorded every second or every hour, but your stakeholders care about the monthly total. This is where resampling becomes essential. It allows you to change the frequency of your data—for example, converting daily sales into monthly revenue—and define exactly how to aggregate those values, whether by summing them, taking the average, or finding the maximum.

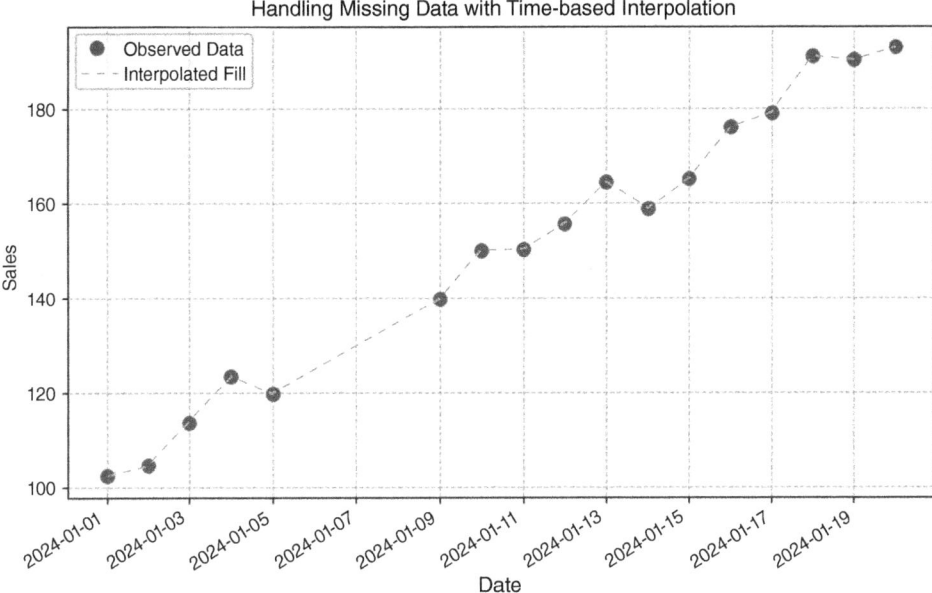

FIGURE 9-2: Handling missing data with time-based interpolation.

This is distinct from the rolling averages used earlier. A rolling average smooths the noise while keeping the original frequency (e.g., daily), whereas resampling fundamentally changes the shape of the data, condensing it into a new timeframe that aligns with your business reporting cycles. In Listing 9-7, daily data is resampled into monthly totals.

LISTING 9-7: RESAMPLING DAILY DATA TO MONTHLY TOTALS

```
# Make sure to run the previous listings before running this code
# Resample 'D' (Daily) data to 'ME' (Monthly) and calculate the Sum
monthly_sales = clean_data.resample('ME').sum()

print("\n--- Monthly Aggregates ---")
print(monthly_sales.head())
```

Listing 9-7 uses the `.resample('ME')` method. This tells Pandas to group our daily data into bins of months. This setting can be set to your data, for example, `W` for week. We then chain `.sum()` to tell it how to handle the data in those bins: add it all up. This transforms our 20 rows of daily data into a single row representing the total sales for January. The result of running this code is as follows:

```
--- Monthly Aggregates ---
                 Sales
2024-01-31  2967.030383
```

However, .sum() is just one of many aggregation methods available. Depending on your business question, you might choose differently:

➤ **.mean():** Useful for finding the average daily performance over a month (e.g., What was our average daily active user count in March?).

➤ **.max() or .min():** Essential for capacity planning (e.g., What was the peak server load last week? or What was the lowest inventory level this quarter?).

➤ **.last():** Critical for financial data like stock prices or account balances, where you care about the value at the end of the period, not the sum or average.

Choosing the right aggregation function is as important as choosing the right timeframe; summing stock prices would be nonsensical, while averaging total revenue would be misleading.

Boxplots and Histograms

While a line chart shows you when things happened, it can obscure what actually happened in terms of variability. To understand the behavior of your metric, you need to ignore time for a moment and look at the distribution. This answers critical questions such as: Is our server load usually stable at around 50%, or does it swing wildly between 10% and 90%?

Two visualizations are particularly powerful here. The *histogram* (often combined with a density plot) shows the overall shape of your data, whether it follows a normal bell curve or is skewed by extreme values. The *boxplot* is excellent for identifying outliers and visualizing the spread of data across different categories, such as comparing the volatility of sales in January versus July. Listing 9-8 creates this visualization.

LISTING 9-8: VISUALIZING TIME-SERIES DISTRIBUTIONS

```
import seaborn as sns

# Generate more data for a better distribution visual
np.random.seed(101)
long_dates = pd.date_range(start='2023-01-01', periods=365, freq='D')
# Data with a trend and some seasonality
daily_volatility = np.random.normal(0, 10, 365)
long_ts = pd.DataFrame({'Value': 100 + daily_volatility}, index=long_dates)

# --- Visualization ---
fig, (ax1, ax2) = plt.subplots(1, 2, figsize=(14, 6))

# 1. Histogram with Density (KDE)
sns.histplot(long_ts['Value'], kde=True, ax=ax1, color='teal')
ax1.set_title('Histogram & Density Plot')
ax1.set_xlabel('Daily Value')

# 2. Boxplot by Month (Checking for Seasonal Volatility)
# We add a 'Month' column for grouping
long_ts['Month'] = long_ts.index.month_name().str[:3] # Jan, Feb, Mar...
sns.boxplot(x='Month', y='Value', data=long_ts, ax=ax2, palette="Blues")
ax2.set_title('Value Distribution by Month')
```

```
plt.tight_layout()
plt.show()
```

In Listing 9-8, the `seaborn` library is used. This library builds on top of Matplotlib and makes creating statistical plots much easier. We use `sns.histplot` to create the histogram on the left, adding a `kde=True` (kernel density estimate) line to smooth out the shape. For the right chart, we use `sns.boxplot`. By setting `x='Month'`, we slice our time-series data into 12 separate buckets, as shown in Figure 9-3. This visualization allows us to instantly compare the volatility of different months side-by-side, revealing seasonal patterns in variance that a simple line chart would hide.

Seasonality and Autocorrelation

In the previous section, we used a rolling average to smooth out short-term fluctuations so we could see the long-term trend. For many businesses, those fluctuations aren't just noise, they are critical patterns. A retailer needs to know exactly how much of their December revenue is due to true growth versus just the holiday rush.

To understand these patterns deeply, we need to look at the memory of our time series. Does high sales today predict high sales tomorrow? Does a spike in January always lead to a slump in February? This is the domain of autocorrelation.

Autocorrelation and Partial Autocorrelation

Autocorrelation function (ACF) is the correlation of a time series with itself at previous time steps, known as *lags*. Think of it as a measure of the echo in your data. If you shout in a canyon, you hear your voice bounce back a few seconds later. In time-series data, an event today (like a hot summer day) might "echo" into tomorrow's sales. ACF answers the question: How much does the value at time t depend on the value at time $t - 1$, $t - 2$, and so on?

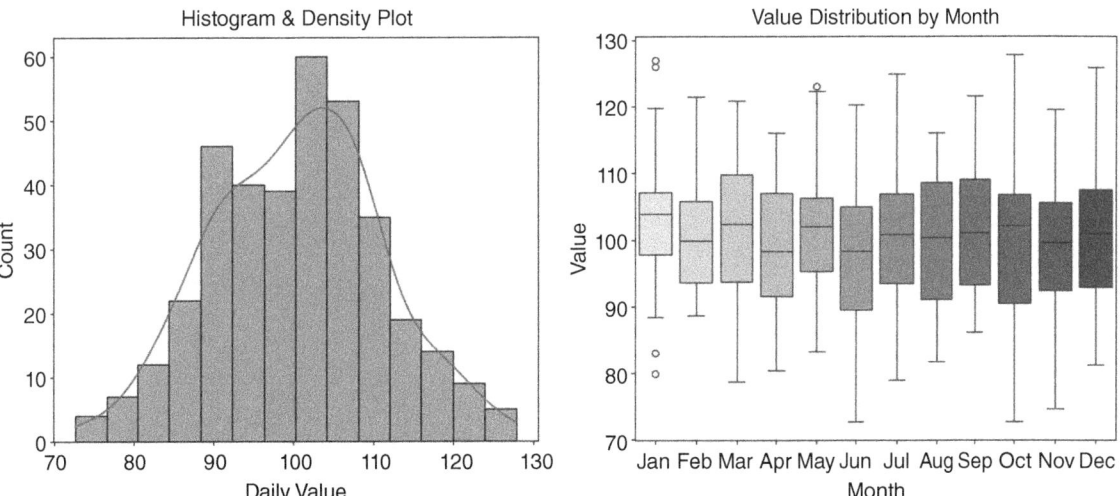

FIGURE 9-3: Histogram and density plot and value distribution by month.

High autocorrelation means the past strongly predicts the future. If today's sales are high, tomorrow's sales are also likely to be high. The echo is loud. *Seasonal autocorrelation* is a specific type of echo that repeats. You might see a strong correlation every seven days (a weekly cycle where Mondays look like other Mondays) or every four quarters (an annual cycle where Q4 always spikes). *Partial autocorrelation* (PACF) is a slightly more sophisticated tool that helps us isolate the direct cause of a correlation. It measures the correlation between today and a past lag, but crucially, it removes the influence of all the steps in between.

In the real world, data is rarely influenced by a single cycle. Most economic datasets exhibit *multipe-riodicity*, where multiple echoes overlap simultaneously. For example, a retail store might see a daily cycle (evenings are busier than mornings), a weekly cycle (Saturdays outperform Tuesdays), and an annual cycle (the December holiday rush). When we plot these correlations, we aren't just looking for one spike; we are looking for a complex interference pattern of these different waves. Recognizing multiperiodicity is crucial because it prevents us from misidentifying a short-term weekly spike as a long-term trend.

To see how PACF helps isolate these layers, imagine a heatwave that affects sales. Imagine a five-day scorcher:

Days 1 and 2: Sales are high due to the onset of the heat.

Day 3: Sales remain high, but is this because of Day 1 or simply because Day 2 was hot?

Days 4 and 5: The trend continues.

If we use standard autocorrelation (ACF), Day 1 will appear highly predictive of Day 5. However, this is a false direct link caused by the intervening days (the ripple effect). PACF mathematically controls for Days 2, 3, and 4. It asks: Once we account for the fact that yesterday was hot, does the temperature from four days ago add any new information? If the answer is no, the PACF for that lag will drop to zero. This allows us to ignore the noise of the heatwave and identify the true seasonal lag, perhaps a recurring weekly delivery day, that actually dictates the inventory we need to stock.

To demonstrate this, Listing 9-9 uses a classic economic dataset: Real Personal Consumption Expenditures (PCE). This data tracks how much money American households spend on goods and services, adjusted for inflation. Crucially, we are using the non-seasonally adjusted version. This means the raw data still contains all the natural spikes and drops of the calendar year.

LISTING 9-9: VISUALIZING AUTOCORRELATION IN ECONOMIC DATA

```
import pandas as pd
import matplotlib.pyplot as plt
from statsmodels.graphics.tsaplots import plot_acf, plot_pacf

# 1. Load the Real Economic Data
# Real Personal Consumption Expenditures (Quarterly)
url = "https://github.com/bkrayfield/Applied-Math-With-Python/raw/refs/heads/main/
Data/ND000349Q.csv"
pce_data = pd.read_csv(url, parse_dates=['observation_date'],
index_col='observation_date')
```

```
# 2. Visualization: ACF and PACF
fig, (ax1, ax2) = plt.subplots(2, 1, figsize=(12, 8))

# Plot Autocorrelation (ACF)
# We look at 20 lags (5 years of quarterly data)
plot_acf(pce_data['ND000349Q'], lags=20, ax=ax1)
ax1.set_title('Autocorrelation Function (ACF) - The "Echo"')

# Plot Partial Autocorrelation (PACF)
plot_pacf(pce_data['ND000349Q'], lags=20, ax=ax2)
ax2.set_title('Partial Autocorrelation Function (PACF) - The "Direct Link"')

plt.tight_layout()
plt.show()
```

Listing 9-9 uses `statsmodels` to generate the ACF and PACF plots. We load the PCE dataset, ensuring the dates are parsed correctly. Then, we create a figure with two subplots. The `plot_acf` function calculates the correlation between the time series and its lagged versions for up to 20 quarters (five years). Similarly, `plot_pacf` calculates the partial autocorrelation. You can see the results of this listing in Figure 9-4.

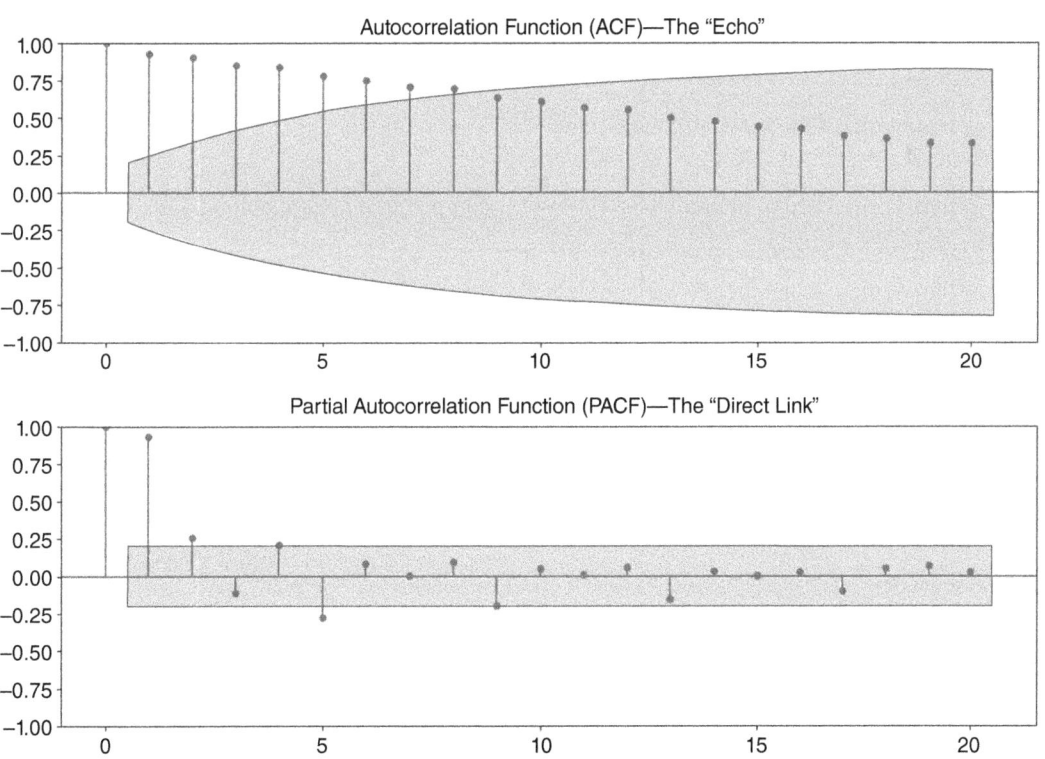

FIGURE 9-4: Autocorrelation and partial autocorrelation plots.

The top chart in Figure 9-4 (ACF) shows a very slow decay. The bars remain high and positive for many lags, indicating a strong long-term trend, values don't change much from one quarter to the next. The bottom chart (PACF) tells a sharper story. The massive spike at Lag 1 indicates that the single best predictor of this quarter's spending is the immediately preceding quarter.

To elaborate more on why these plots are important, the slow decay in the ACF plot (top chart) is a classic signature of a trend. It tells us that the data points are sticky. For example, if sales were high last quarter, they are likely to be high this quarter, next quarter, and so on. This confirms that the long-term growth we saw in the line chart isn't an illusion; it's a statistically significant property of the data. This plot shows how each progressive data point is linked as a chain. The second PACF plot (bottom plot) looks at how specific historical data points affect what happens to future data points. For example, if the line was larger four quarters ago, then we can conclude what happened one year ago has a direct influence on what happens today.

In short, these plots mathematically demonstrate why we need to use a model that accounts for both trend and seasonality (like `seasonal_decompose` or SARIMA), rather than a simple average. They move you from guessing there's a pattern to knowing exactly what that pattern looks like.

Time-series Decomposition

The previous section used a rolling average to smooth out short-term fluctuations so we could see the long-term trend. For many businesses, those fluctuations aren't just noise, they are critical patterns. A retailer needs to know exactly how much of their December revenue is due to true growth versus just the holiday rush.

To answer this, we use time-series decomposition. This statistical technique takes a single line chart and mathematically splits it into three distinct components:

➤ **Trend:** The long-term direction (up or down), stripped of all other noise.

➤ **Seasonality:** The repeating patterns that happen at fixed intervals (e.g., every December, every weekend).

➤ **Residuals (noise):** The random randomness that remains after the trend and seasonality are removed. This is the "unexplained" part of the data.

To demonstrate this, we'll continue with the PCE dataset. Even more important, we are using the non-seasonally adjusted version. This means the raw data still contains all the natural spikes and drops of the calendar year (like the massive surge in spending every Q4 for the holidays). This makes it the perfect candidate for decomposition: we can use Python to find that holiday signal and separate it from the underlying economic growth.

In this analysis, we use a multiplicative model. Unlike an additive model, which assumes seasonal swings are a constant dollar amount, a multiplicative model assumes that seasonal effects are proportional to the trend. As the economy grows, the holiday spike scales up with it. Mathematically, we are representing the data as follows:

$$Observed = Trend * Seasonality * Residual$$

We can perform this advanced analysis with a single function from the statsmodels library called seasonal_decompose, shown in Listing 9-10.

LISTING 9-10: VISUALIZING SEASONALITY VS. TREND IN ECONOMIC DATA

```
import pandas as pd
import matplotlib.pyplot as plt
from statsmodels.tsa.seasonal import seasonal_decompose

# 1. Load the Real Economic Data
# Real Personal Consumption Expenditures (Quarterly)
url = "https://github.com/bkrayfield/Applied-Math-With-Python/raw/refs/heads/main/
Data/ND000349Q.csv"
pce_data = pd.read_csv(url, parse_dates=['observation_date'],
index_col='observation_date')

# 2. Run the Decomposition
# We specify 'period=4' because the data is Quarterly (4 points per year)
# model='multiplicative' is often better for economic data where volatility
grows with the trend,
# but 'additive' is easier to interpret for a first example.
result = seasonal_decompose(pce_data['ND000349Q'], model=' multiplicative',
period=4)

# 3. Visualization
# The result object has a built-in .plot() function that creates a 4-panel chart
fig = result.plot()
fig.set_size_inches(10, 8) # Make it large enough to read
plt.suptitle('Decomposing Personal Consumption Expenditures (PCE)', fontsize=16,
y=1.02)
plt.show()
```

In Listing 9-10, the core analysis happens with the seasonal_decompose function. We pass it two critical arguments. First, model='multiplicative' tells the function to treat the seasonal component as a percentage or factor of the trend. This is a more realistic x-ray for the PCE data because, over several decades, U.S. spending has increased significantly; it makes sense that the holiday surge in 2023 is much larger in absolute dollars than it was in 1970, even if the percentage of growth remains similar. Second, period=4 informs the algorithm that our data is quarterly, identifying the pattern that repeats every four observations.

The resulting plot in Figure 9-5 reveals the distinct layers hidden within the raw data. The top panel, Observed, shows the raw, jagged path of the economy. The second panel, Trend, reveals the smooth underlying growth, clearly highlighting structural shifts like the flattening during the 2008 financial crisis.

The third panel, Seasonal, is where the multiplicative logic shines. Instead of showing dollar amounts, it shows a ratio. A value of 1.05 in Q4 would indicate that spending is 5% higher than the trend due to the season. This consumer heartbeat remains consistent even as the economy scales. Finally, the Residuals panel shows the noise. Because this is a multiplicative model, the residuals are centered on 1.0 rather than 0. Any sharp deviations from 1.0 represent shocks to the system, unpredictable events like the 2020 pandemic lockdowns that disrupted both the trend and the expected seasonal rhythm.

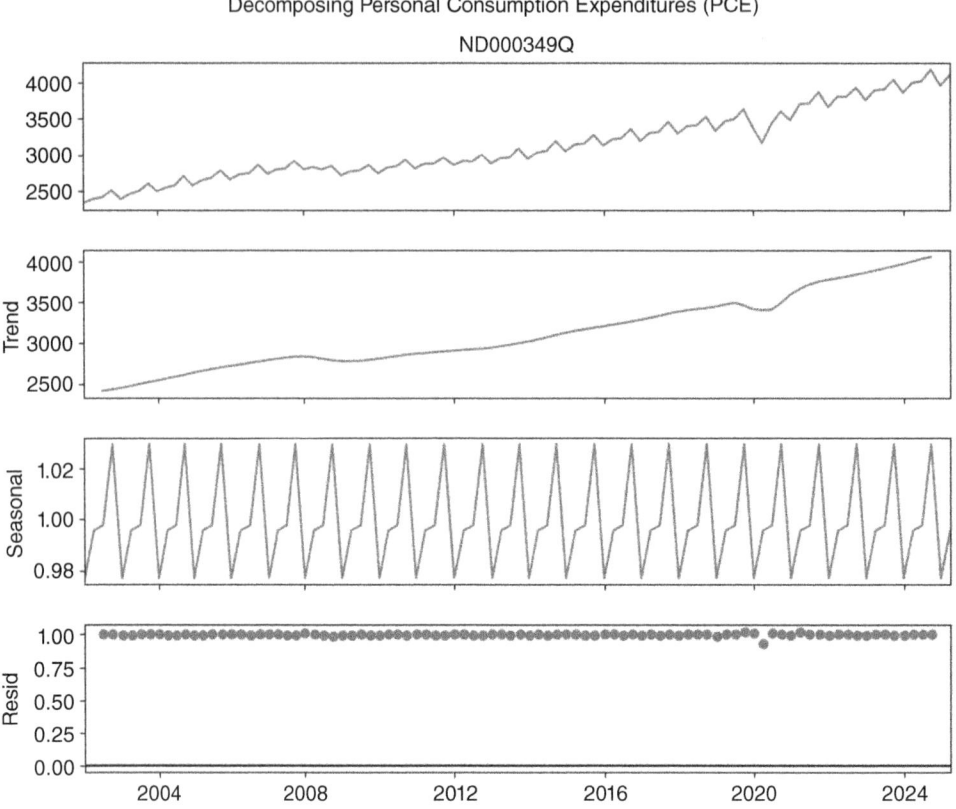

FIGURE 9-5: A time-series decomposition of U.S. consumer spending.

PANEL DATA

Finally, we turn to the most complex data structure: panel data. This is data that tracks multiple subjects over multiple time periods. Imagine a dataset tracking the price of energy across different sectors (residential, commercial, industrial) and fuel types over 50 years.

If you try to plot this on a single line chart, you get a spaghetti chart, a tangled mess of overlapping lines that is impossible to read. The solution is small multiples (or faceting). Instead of one big chart, you create a grid of smaller charts, one for each category. This allows the eye to easily compare trends without visual clutter.

Listing 9-11 creates small multiples using data composed of energy prices by sectors. We use a dataset of energy prices in New York State from 1970 to 2022. It contains prices for different fuel types (natural gas, electricity) across different economic sectors.

LISTING 9-11: CREATING SMALL MULTIPLES WITH SEABORN

```
import pandas as pd
import matplotlib.pyplot as plt
import seaborn as sns

# 1. Load the Raw Data
url = "https://github.com/bkrayfield/Applied-Math-With-Python/raw/refs/heads/main/
Data/Energy_Prices,_Dollars_per_Million_Btu__Beginning_1970_20251201.csv"
energy_data = pd.read_csv(url)

# 2. Reshape from "Wide" to "Long" Format
# The data has fuel types as separate columns (Coal, Natural Gas, etc.).
# Seaborn prefers "Long" format: one column for "Fuel Type" and one for "Price".
fuel_cols = ['Coal', 'Propane', 'Natural Gas', 'Electricity']

energy_long = energy_data.melt(
    id_vars=['Year', 'Sector'], # Identifiers to keep
    value_vars=fuel_cols,       # Columns to unpivot
    var_name='Fuel Type',       # New column name for headers
    value_name='Price'          # New column name for values
)

# 3. Clean the Price Column
# Prices have '$' signs (e.g., "$1.09"), so pandas reads them as strings/objects.
# We must remove the '$' and convert to float.
energy_long['Price'] = energy_long['Price'].astype(str).str.
replace('$', '', regex=False)
energy_long['Price'] = pd.to_numeric(energy_long['Price'], errors='coerce')

# Filter for just "Residential" to make the chart focused
residential_energy = energy_long[energy_long['Sector'] == 'Residential']

# 4. Create Small Multiples
g = sns.relplot(
    data=residential_energy,
    x="Year", y="Price",
    col="Fuel Type", # Create a separate chart for each Fuel Type
    kind="line",
    col_wrap=2,      # Start a new row after 3 charts
    height=2, aspect=1.5,
    linewidth=2
)

# Add Titles
g.fig.suptitle('Residential Energy Price Trends (Per Million BTU1970-2021)',
y=1.02, fontsize=16)
plt.show()
```

In Listing 9-11, we use seaborn's relplot function, which is designed specifically for handling complex, multi-dimensional data. The key argument is col="Fuel Type". This tells seaborn to slice the data by fuel type and automatically generate a separate subplot for each one. The col_wrap=2 code

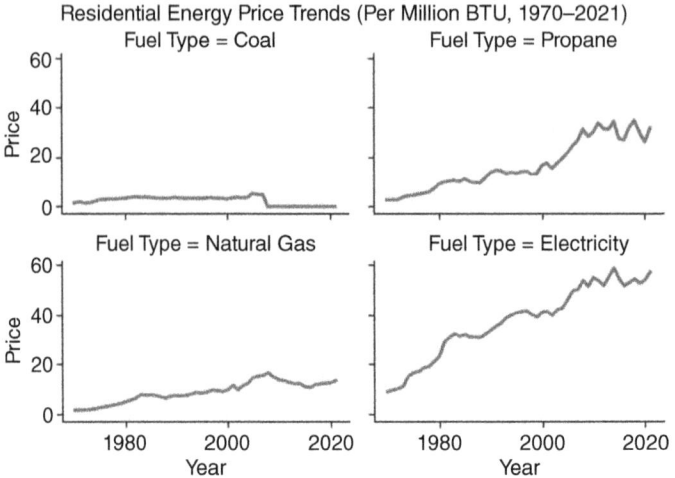

FIGURE 9-6: Residential energy prices.

ensures the charts are arranged in a neat grid rather than in one long row. You can see this visual in Figure 9-6.

This visualization technique reveals insights that would be hidden in a single chart. We can instantly see that electricity prices have been relatively stable and high, while fuel oil and propane prices are extremely volatile, spiking dramatically during geopolitical crises. By separating the signals, we respect the complexity of the data while keeping the story clear.

SUMMARY

This chapter explored the critical role of data visualization in translating raw numbers into actionable business insights. It established that the first step in any visualization task is diagnosing the fundamental structure of the data: cross-sectional (a snapshot in time), time-series (a historical sequence), or panel data (multidimensional history). The chapter demonstrated how identifying these structures allows analysts to select the appropriate visualization strategy, whether utilizing bar charts for direct comparisons, line charts for analyzing trends, or faceted plots to untangle complex, multi-subject datasets.

It then focused heavily on the mechanics of working with time-series data, utilizing Python's Matplotlib and Pandas libraries to move beyond simple plotting. You learned about techniques for revealing long-term signals amid short-term noise using rolling averages and addressed common real-world data issues by applying time-based interpolation to repair missing values. The chapter also examined how to change the temporal resolution of data through resampling, aggregating granular observations into meaningful business cycles, and how to visualize volatility and distribution using histograms and boxplots.

Finally, the chapter introduced advanced diagnostic techniques to mathematically quantify the patterns hidden within time-series data. It utilized autocorrelation (ACF) and partial autocorrelation

(PACF) to measure the memory or echo within a dataset and applied time-series decomposition to separate data into its constituent parts: trend, seasonality, and residual noise. The chapter concluded by applying these principles to panel data, using `seaborn` to create organized, comparative views that prevent visual clutter and highlight relationships across different categories over time.

CONTINUE YOUR LEARNING

Data visualization and time-series analysis are disciplines that bridge the gap between engineering and art. While this chapter provided the technical foundation for revealing patterns in noise, mastering the aesthetic and theoretical sides of these topics will make your analysis significantly more persuasive. The following resources are curated to help you master the libraries we used and deepen your theoretical understanding:

Official Documentation

➤ **Matplotlib:** The foundational library for Python plotting. The gallery section is particularly useful for finding code snippets for specific chart types. `https://matplotlib.org/stable/gallery/index.html`

➤ **Seaborn:** The high-level interface for statistical graphics. Their tutorial on visualizing statistical relationships is excellent for understanding panel data and complex distributions. `https://seaborn.pydata.org/tutorial.html`

➤ **Pandas Time Series:** The definitive guide to the functionality that makes Python so powerful for financial and economic analysis, covering offsets, shifting, and frequency conversion. `https://pandas.pydata.org/docs/user_guide/timeseries.html`

➤ **Statsmodels Time Series analysis:** Deep documentation for the advanced components we touched on, such as decomposition, stationarity tests, and autocorrelation. `https://www.statsmodels.org/stable/tsa.html`

Recommended Reading

➤ *Storytelling with Data* by **Cole Nussbaumer Knaflic:** This book focuses less on code and more on the design principles for creating charts that effectively communicate a message to stakeholders. It is essential reading for the "last mile" of analytics.

➤ *Python for Data Analysis* by **Wes McKinney:** Written by the creator of Pandas, this book offers the most in-depth look at the mechanics of data manipulation, particularly for time-series operations and cleaning messy datasets.

➤ *Effective Pandas 2* by **Matt Harrison:** A standard in the field for learning idiomatic Pandas and data manipulation patterns. This is an excellent resource for those looking to write "Treading on Python" style code, with updated editions covering the latest features in Pandas.

➤ *Forecasting: Principles and Practice* by **Hyndman and Athanasopoulos:** While the code examples are in R, this is widely considered the gold standard textbook for the theory behind time-series decomposition and forecasting.

In addition to generating charts, you will frequently need to manipulate the shape of your time-series data to prepare it for analysis. Table 9-1 serves as a quick reference for the essential methods in Pandas and Statsmodels used to structure, smooth, and diagnose temporal data.

TABLE 9-1: Essential Visualization and Time-series Functions

FUNCTION/METHOD	LIBRARY	DESCRIPTION
`.to_datetime()`	Pandas	Converts string arguments to datetime objects. The critical first step for any time-series analysis.
`.set_index()`	Pandas	Moves a column (usually the date) to the DataFrame index, enabling time-aware slicing and plotting.
`.rolling(window=n).mean()`	Pandas	Calculates a moving average over a specified window n to smooth out short-term noise and reveal trends.
`.interpolate(method='time')`	Pandas	Fills missing values (NaN) by drawing a line between existing points, respecting the time distance between them.
`.resample(rule).func()`	Pandas	Changes the frequency of the data (e.g., Daily to Monthly). Must be chained with an aggregation function like `.sum()` or `.mean()`.
`.shift(periods=n)`	Pandas	Shifts the index by n periods. Essential for calculating percent changes or creating lag features for models.
`seasonal_decompose()`	Statsmodels	Mathematically separates a time series into three distinct components: Trend, Seasonality, and Residuals.
`plot_acf() / plot_pacf()`	Statsmodels	Visualizes autocorrelation and partial autocorrelation to diagnose the "memory" and cyclic dependency in the data.
`sns.relplot(kind='line')`	Seaborn	The primary function for creating "small multiples" (faceted plots) to visualize panel data without clutter.

10

Illustrating Cross-sectional Data

If time-series analysis, covered in the previous chapter, is akin to watching a movie of your business history, cross-sectional analysis is like examining a high-resolution photograph. You are freezing time to look at the relationships between different entities at a single, distinct moment. In the previous chapter, we asked how we got here. This chapter pivots to an equally critical question—where are we right now?

Cross-sectional visualization allows you to ignore the timeline and focus on structure. It answers questions of rank, such as identifying which product is the current bestseller; questions of distribution, such as determining if your customer base is predominantly young or old; and questions of correlation, such as asking if higher advertising spend actually leads to higher sales volume. In this chapter, you will master the art of comparison using bar charts, explore the shape of data using histograms and boxplots, and reveal hidden relationships between variables using scatterplots.

DATA CATEGORIES

Business questions often revolve around structure rather than magnitude. We need to understand the makeup of our data. This is the domain of categorical analysis, specifically part-to-whole comparisons. Whether you are breaking down a marketing budget, analyzing market share, or looking at the inventory mix of a warehouse, the goal is to visualize how multiple small parts combine to form the total picture. The following sections explore how to answer these questions using visuals.

The Pie Chart

When the analytical question shifts from ranking to composition, we are no longer looking for the highest value; we are looking for the share of the total. We want to know how a specific entity, like a budget, a market, or a dataset, is divided into its constituent parts. The fundamental tool for this is the pie chart.

The pie chart represents a single categorical variable as a circle, where the entire area corresponds to 100% of the data. The circle is sliced into sectors, with the arc length and angle of each slice proportional to the category's contribution to the whole. It provides stakeholders with an immediate, intuitive sense of proportion, allowing them to quickly identify dominant categories without reading specific numbers.

Listing 10-1 visualizes the composition of our vegetable dataset (from Chapter 9) to see the breakdown of different forms (Fresh, Canned, Frozen, etc.). As a reminder, the vegetable dataset is a cross-sectional dataset showing the prices of different vegetables and the manner they are delivered (form: like Frozen or Fresh).

> **NOTE** Note that Listing 10-1 uses a CSV file called `Vegetable-Prices-2022.csv` found on GitHub. This file, along with the files used in the other listings in this chapter, are also included in the downloadable files for this book, located on the Wiley site at `www.wiley.com/go/PythonMath`.

LISTING 10-1: THE STANDARD PIE CHART

```python
import matplotlib.pyplot as plt
import pandas as pd
import seaborn as sns

# Load the Data
url = "https://github.com/bkrayfield/Applied-Math-With-Python/raw/refs/heads/main/
Data/Vegetable-Prices-2022.csv"
veg_prices = pd.read_csv(url)

# 1. Prepare the Data
# We count the frequency of each form to get the 'parts' of the whole
form_counts = veg_prices['Form'].value_counts()

# 2. Visualization
plt.figure(figsize=(8, 8))

# We use Matplotlib's native pie function
plt.pie(
    form_counts,
    labels=form_counts.index,
    autopct='%1.1f%%',      # Format the percentages (e.g., 15.5%)
    startangle=140,         # Rotate the chart to a pleasing angle
    colors=sns.color_palette('pastel') # Use a soft color palette
)

plt.title('Composition of Vegetable Forms', fontsize=16)
plt.show()
```

In Listing 10-1, we utilize Matplotlib's `.pie()` function. Unlike bar or scatterplots, which require X and Y coordinates, a pie chart requires only a single array of numerical values (`form_counts`). The function automatically calculates the total sum and determines the angle for each slice. We utilize the `autopct` parameter to overlay the calculated percentages directly onto the chart; the string format `'%1.1f%%'` instructs Python to display the number as a float with one decimal place followed by a percent sign. Finally, `startangle=140` allows us to rotate the entire chart, ensuring that the labels are positioned in the most readable orientation.

The resulting plot is shown in Figure 10-1. This figure utilizes a pie chart to visualize the composition of the dataset by vegetable form, providing an immediate "part-to-whole" comparison. The chart reveals that fresh vegetables dominate the dataset, accounting for nearly half of all items at 45.2%. The remaining half is split between processed forms, with canned vegetables representing 25.8%, frozen vegetables representing 20.4%, and dried vegetables making up the smallest portion at 8.6%. This distribution clearly indicates that the dataset is balanced roughly 50/50 between fresh produce and preserved alternatives.

We can further customize the pie chart to emphasize specific insights or improve aesthetics. If a particular category requires immediate attention, such as highlighting the prevalence of fresh vegetables, we can use the `explode` parameter. This argument accepts a collection of values corresponding to the slices; setting a non-zero value (e.g., 0.1) for a specific slice will "explode" or offset it from the center, isolating it visually. Additionally, analysts often prefer a donut chart variation, which can be achieved in Matplotlib by adding the `wedgeprops` argument (e.g., `wedgeprops={'width': 0.5}`). This hollows out the center, reducing the visual mass of the chart and shifting the focus to the length of the arcs rather than the total area.

Donut Charts

In the business world, the pie chart is ubiquitous. It is the default choice for showing part-to-whole composition, such as market share or budget allocation. However, among data scientists and

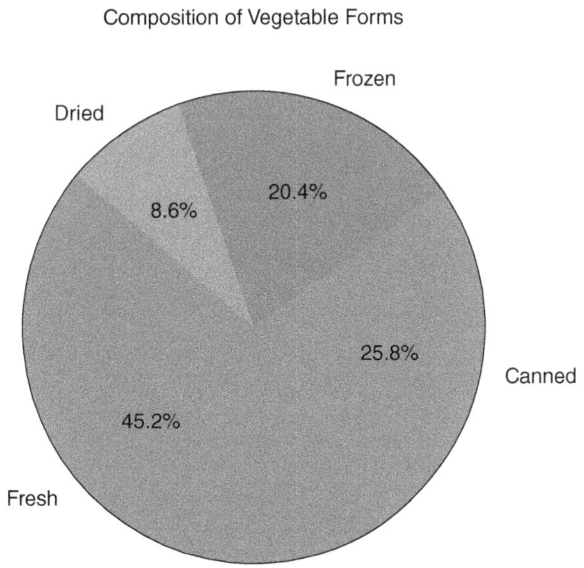

FIGURE 10-1: Pie chart of vegetables.

visualization experts, it is viewed with skepticism. The criticism is not just aesthetic; it is functional. Pie charts are notoriously difficult to read with precision. When slices are similar in size, it is nearly impossible for a viewer to distinguish the difference based on the angle alone. Furthermore, comparing slices that are not adjacent requires the viewer to mentally rotate the shapes, increasing cognitive load and the likelihood of error.

Despite these limitations, stakeholders often demand them because they provide an immediate, intuitive sense of "wholeness" that a bar chart lacks. If you must use a circular visualization, the donut chart is a superior alternative. By removing the center, you remove the most difficult aspect of the chart to interpret, the angles at the vertex. This forces the eye to compare the arc lengths of the outer ring, which is slightly more intuitive. Additionally, the empty center provides valuable real estate to display a summary statistic, such as the grand total, making the chart more information-dense.

Listing 10-2 visualizes the composition of our vegetable dataset to show the proportion of items that are fresh versus canned versus frozen.

LISTING 10-2: CREATING A DONUT CHART WITH MATPLOTLIB

```
import matplotlib.pyplot as plt
import pandas as pd

# Load the Data
url = "https://github.com/bkrayfield/Applied-Math-With-Python/raw/refs/heads/main/
Data/Vegetable-Prices-2022.csv"
veg_prices = pd.read_csv(url)

# 1. Prepare the Data
# Count the frequency of each form
form_counts = veg_prices['Form'].value_counts()

# 2. Visualization
plt.figure(figsize=(8, 8))

# Create the Pie Chart
# autopct formats the values as percentages (e.g., '12.5%')
# startangle=90 rotates the start to the top (12 o'clock)
plt.pie(
    form_counts,
    labels=form_counts.index,
    autopct='%1.1f%%',
    startangle=90,
    colors=sns.color_palette('pastel'),
    wedgeprops={'edgecolor': 'white', 'linewidth': 2}
)

# 3. Transform into a Donut
# We draw a white circle in the center to cover the middle
centre_circle = plt.Circle((0, 0), 0.70, fc='white')
fig = plt.gcf()
fig.gca().add_artist(centre_circle)
```

```
plt.title('Distribution of Vegetable Forms in Dataset', fontsize=16)
plt.show()
```

In Listing 10-2, we rely on Matplotlib's foundational `.pie()` function. The transformation into a donut is a visual hack. Matplotlib does not have a native donut function. Instead, we create a standard pie chart and then instantiate a `plt.Circle` object. The arguments `(0, 0)` and `0.70` place the circle at the origin with a radius of 0.7 (covering 70% of the pie). We then access the current figure using `plt.gcf()` and "add the artist" (the circle) on top of the existing plot. This technique creates the modern ring aesthetic while preserving the underlying statistical proportions.

The `.pie()` function offers several parameters to refine the presentation of categorical data beyond basic slices. You can use `explode` to pass an array of offsets that pull specific slices away from the center, which is ideal for highlighting a particular data point. The `wedgeprops` dictionary allows for fine-grained control over the slices themselves, such as setting `edgecolor` and `linewidth` to create clear boundaries or using width to transform the pie into a donut chart. Additionally, `textprops` can be used to modify the font size and color of labels, while the `normalize` parameter ensures that the data scales correctly to a full circle even if the input values don't sum to 1 or 100.

Figure 10-2 presents the same composition data as the previous pie chart but utilizes a donut chart format. By removing the center, the visualization shifts the viewer's focus from the angles at the vertex to the arc lengths of the outer ring, which many find easier to compare. The statistical breakdown remains identical; fresh vegetables comprise the clear majority at 45.2%, followed by canned at 25.8%, frozen at 20.4%, and dried at 8.6%, but the inclusion of whitespace in the middle produces a cleaner aesthetic and eliminates the visual clutter where the slices would normally converge.

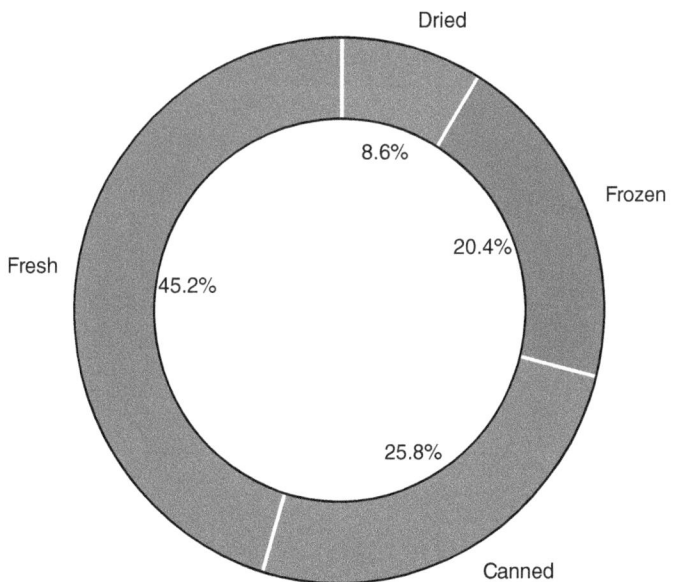

FIGURE 10-2: Donut plot.

The `.pie()` function offers several parameters to further refine the presentation of categorical data beyond basic slices. You can use the `explode` parameter to pass an array of offsets that pull specific slices away from the center, which is ideal for highlighting a particular data point like the dominant Fresh category. The `wedgeprops` dictionary allows for fine-grained control over the slices themselves; beyond setting the `edgecolor` and linewidth to create clear boundaries, you can also use the width key here to create a donut chart natively (without the circle overlay hack). Additionally, `textprops` can be used to modify the font size and color of labels, while the `normalize` parameter ensures that the data scales correctly to a full circle even if the input values don't sum to exactly 1 or 100.

Stacked Bar Charts

While pie and donut charts are effective for high-level summaries, they can be inefficient when space is limited or when you need to compare multiple compositions side-by-side. In these instances, the stacked bar chart offers a distinct advantage. It functions effectively as a linear pie chart, unrolling the circle into a single rectangular bar.

This transformation allows viewers to judge proportions based on length, a task the human eye performs with high accuracy, rather than angle or area. Furthermore, stacked bars are exceptionally space-efficient. They allow you to display complex part-to-whole relationships within the compact rows of a table or a tight dashboard panel, where a circular chart would be too bulky to be legible.

Listing 10-3 utilizes Matplotlib's core primitives to build a custom visualization. This is necessary because a stacked bar is technically just a series of standard bars placed end-to-end.

LISTING 10-3: THE STACKED BAR (A BETTER ALTERNATIVE)

```python
import matplotlib.pyplot as plt
import pandas as pd

# Load the Data
url = "https://github.com/bkrayfield/Applied-Math-With-Python/raw/refs/heads/main/
Data/Vegetable-Prices-2022.csv"
veg_prices = pd.read_csv(url)

# 1. Create a Frequency Table (Composition Data)
# We count how many vegetables exist for each form
composition = veg_prices['Form'].value_counts().reset_index()
composition.columns = ['Form', 'Count']

# Calculate Percentage
total = composition['Count'].sum()
composition['Percentage'] = (composition['Count'] / total) * 100

# 2. Visualization
plt.figure(figsize=(10, 2))

# Create a horizontal stacked bar
# We use the 'Percentage' column for the width so the axis spans 0-100
left = 0
```

```
for i, row in composition.iterrows():
    plt.barh(
        y=0,
        width=row['Percentage'],
        left=left,
        label=f"{row['Form']} ({row['Percentage']:.0f}%)"
    )
    left += row['Percentage']

plt.title('Dataset Composition: Vegetable Forms by Percentage')
plt.yticks([]) # Remove y-axis ticks
plt.xlabel('Percentage of Total (%)')
plt.xlim(0, 100) # Force the axis to end exactly at 100

# Adjust legend placement
# bbox_to_anchor moves the legend relative to the anchor point
plt.legend(ncol=4, loc='upper center', bbox_to_anchor=(0.5, -0.35), frameon=False)

# Explicitly add space at the bottom for the legend
plt.subplots_adjust(bottom=0.45)

plt.show()
```

For this stacked bar chart, we manage the placement of the bars using the `left` variable, which acts as an accumulator. It starts at 0 and increments by the width of each bar (`left += row['Percentage']`) after every iteration of the loop. This ensures that the start of the next bar aligns perfectly with the end of the previous one.

Critically, Listing 10-3 visualizes the percentage of vegetable forms (Canned, Fresh, or Frozen), not the raw count. By setting `width=row['Percentage']` and enforcing `plt.xlim(0, 100)`, we normalize the visual scale, guaranteeing the bar spans exactly from 0 to 100 units regardless of the dataset size.

Finally, we address the common issue of legend overlap using `plt.subplots_adjust(bottom=0.45)`. This command essentially shrinks the height of the chart content within the window, creating a reserved margin of whitespace at the bottom where the legend can reside without obstructing the data or being clipped by the window frame.

The result from this listing is the chart shown in Figure 10-3. This figure visualizes the dataset composition using a horizontal stacked bar chart, essentially unrolling the previous pie chart into a single linear track. The entire length of the bar represents 100% of the data, segmented by color to show the relative contribution of each form. Fresh vegetables clearly dominate the distribution, occupying the first 45% of the bar, followed by canned at 26%, frozen at 20%, and dried at 9%. This layout facilitates a direct comparison of segment lengths along the x-axis, offering a more precise and space-efficient alternative to circular charts for gauging part-to-whole relationships.

CORRELATIONS AND DISTRIBUTIONS

The most common task in business analytics is ranking. You are often presented with a categorical list, sales reps, product lines, or store locations, and the immediate business need is to identify who is

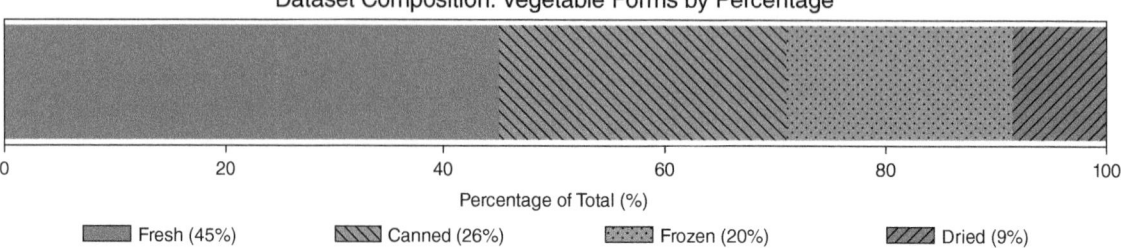

FIGURE 10-3: Stacked bar chart.

outperforming the pack and who is lagging behind. While a spreadsheet or a table offers precision to the umpteenth decimal place, it fails at pattern recognition. To find the maximum value in a table of 50 states, your brain must read 50 individual numbers, hold them in short-term memory, and compare them. A visual comparison makes the maximum obvious in milliseconds. Two ways to do visual comparisons are bar charts and boxplots.

Bar Charts

The bar chart (sometimes called a bar plot) is the undisputed workhorse for this type of cross-sectional comparison. However, a common mistake that analysts make is sticking to the default vertical alignment found in most software. When your category names are long, for example, "Enterprise Software License" versus "Consumer App," vertical labels often overlap, become unreadable, or get rotated 90 degrees, forcing the readers to tilt their heads to read the axis.

To solve this, we apply a simple heuristic: if you have more than five categories, or if your category names are long, use a horizontal bar chart. Furthermore, the order of the bars matters immensely. An unsorted bar chart is just a column forest that forces the eye to jump back and forth to compare heights. Sorted bars create a staircase effect, allowing the viewer to instantly group the high-cost items versus the low-cost items. Another solution to this issue could be to rotate the labels 45 degrees.

Listing 10-4 creates a hierarchal bar chart of the vegetable data we introduced earlier in this chapter to compare retail prices effectively.

LISTING 10-4: THE HORIZONTAL BAR CHART

```python
import pandas as pd
import matplotlib.pyplot as plt
import seaborn as sns

# 1. Load the Data
url = "https://github.com/bkrayfield/Applied-Math-With-Python/raw/refs/heads/main/
Data/Vegetable-Prices-2022.csv"
veg_prices = pd.read_csv(url)

# 2. Prepare the Data
# Filter for just "Fresh" vegetables to make a fair comparison
```

```
# and sort the values to create a logical "staircase" visual
fresh_veg = veg_prices[veg_prices['Form'] == 'Fresh'].sort_values('RetailPrice', as
cending=False).head(10)

# 3. Visualization
plt.figure(figsize=(10, 6))

# We use orient='h' for horizontal bars to accommodate long labels
sns.barplot(
    data=fresh_veg,
    x='RetailPrice',
    y='Vegetable',
    color='seagreen'
)

plt.title('Top 10 Most Expensive Fresh Vegetables (2022)', fontsize=14)
plt.xlabel('Price per Pound ($)')
plt.ylabel('') # Remove the y-label as it's self-explanatory
plt.grid(axis='x', linestyle='--', alpha=0.7)

plt.show()
```

In Listing 10-4, the code begins with a rigorous data preparation phase using Pandas chained operations. We first apply a Boolean mask [veg_prices['Form'] == 'Fresh'] to isolate a specific subset of data; comparing fresh produce to canned or frozen goods would introduce skew due to processing costs, so this filtering is statistically essential. Immediately following the filter, we invoke. sort_values('RetailPrice', ascending=False). This is a critical step for visualization; without sorting the DataFrame before passing it to the plotting library, the resulting chart would display bars in random index order, destroying the ability to quickly rank items. We conclude the chain with. head(10) to limit our dataset to the top outliers, preventing the chart from becoming overcrowded.

The visualization itself relies on Seaborn's sns.barplot() function. We explicitly map the quantitative variable RetailPrice to the x-axis and the categorical variable Vegetable to the y-axis. While modern versions of Seaborn can infer orientation based on data types, explicit mapping ensures stability. By setting color='seagreen', we override the default multi-colored palette; in a ranking chart where distinct colors do not represent distinct data groups, using a single uniform color reduces cognitive load and keeps the focus on the length of the bars. Finally, plt.grid(axis='x') is added to improve readability, allowing the eye to trace the end of a bar down to the specific value on the x-axis.

We see in Figure 10-4 that Okra is the most expensive vegetable in our data, followed by spinach, mushrooms, and more. The bar chart provides a simple way to determine quantity. The choice of vertical or horizontal is usually based on the preference or the shape of your data.

Seaborn's barplot function offers extensive customization options to adapt the chart to more complex data stories. While we used color to set a uniform tone, you can use the hue parameter to introduce a second categorical variable, which will split each bar into sub-groups (comparing prices across different years side-by-side). You can also utilize the palette parameter to apply meaningful color maps, such as a diverging palette if your data centers around zero, or a sequential palette to emphasize magnitude. Additionally, because Seaborn is built on top of Matplotlib, you can fine-tune axes using standard commands; for example, if you prefer vertical bars but have tight spacing, you

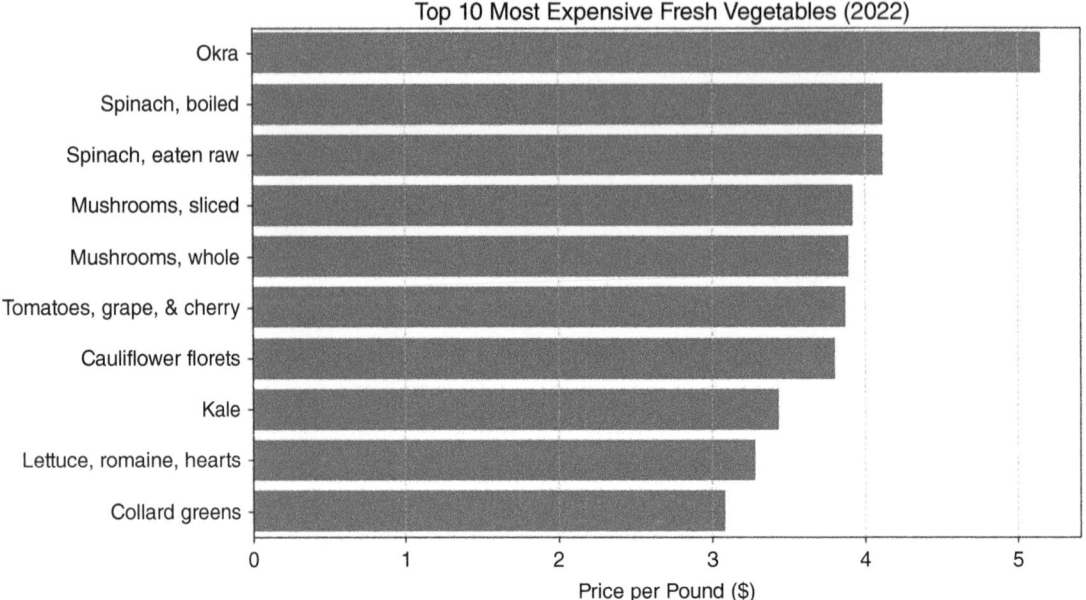

FIGURE 10-4: Bar chart of fresh vegetables.

can use `plt.xticks(rotation=45)` to angle your labels legibly. Finally, the `errorbar` parameter (formerly `ci`) allows you to automatically calculate and display confidence intervals, adding a layer of statistical rigor to your visual comparison.

Boxplots

One of the most misunderstood concepts in data analysis is the average. Averages obscure reality. For example, if you have two people, and one earns nothing and the other earns $100,000, the average salary is $50,000, a number that accurately describes neither person. To truly understand cross-sectional data, you must understand its distribution. You need to know if your data is clustered tightly around the middle (a *normal distribution*) or if it is skewed by a few (a *long-tail distribution*).

For example, we can ask whether fresh produce is significantly more volatile in price than canned produce.

To visualize this in Listing 10-5, we use the boxplot. It is a standardized way of displaying data based on a five-number summary: minimum, first quartile (25%), median, third quartile (75%), and maximum.

LISTING 10-5: COMPARING DISTRIBUTIONS WITH BOXPLOTS

```
import pandas as pd
import matplotlib.pyplot as plt
import seaborn as sns
```

```
# Load the Data
url = "https://github.com/bkrayfield/Applied-Math-With-Python/raw/refs/heads/main/
Data/Vegetable-Prices-2022.csv"
veg_prices = pd.read_csv(url)

plt.figure(figsize=(12, 6))

# We filter out rare forms to keep the chart clean
common_forms = veg_prices[veg_prices['Form'].isin(['Fresh', 'Canned', 'Frozen',
'Dried'])]

sns.boxplot(
    data=common_forms,
    x='Form',
    y='RetailPrice',
    palette='coolwarm'
)

plt.title('Price Distributions by Vegetable Form')
plt.xlabel('Form')
plt.ylabel('Retail Price per Pound ($)')
plt.grid(axis='y', linestyle='--', alpha=0.3)
plt.show()
```

The code in Listing 10-5 delegates significant statistical computation to the `sns.boxplot` function. When we pass `x='Form'` and `y='RetailPrice'`, Seaborn groups the DataFrame by the unique values in the `Form` column. For each group, it automatically calculates the interquartile range (IQR), which is the distance between the 25th percentile and the 75th percentile. This range forms the colored box in the visual (see Figure 10-5). The function then calculates the whiskers (usually 1.5 times the IQR) to determine reasonable boundaries for the data. Any data point existing outside these calculated whiskers is rendered as an individual diamond or dot. This automatic outlier detection is why the boxplot is superior to a bar chart for risk analysis; it visually separates the "normal" variation from the extreme anomalies (like the wildly expensive dried mushrooms) without requiring the user to write manual filtering logic.

Figure 10-5 is the result of Listing 10-5, which utilizes a boxplot to compare the statistical distribution of retail prices across four vegetable forms: fresh, canned, frozen, and dried.

When interpreting Figure 10-5, it is important to understand the specific statistical attributes that define the boxplot's structure. The box represents the interquartile range (IQR), while the horizontal line within it denotes the median. In Seaborn, these elements are controlled by specific parameters: `whis` defines the length of the whiskers (commonly set to 1.5 times the IQR), and `showfliers` determines whether the extreme outliers (the dots seen in the Canned and Frozen categories) are displayed. Furthermore, you can enhance the comparison using the `notch=True` attribute, which creates a narrowed area around the median to represent a confidence interval, or `showmeans=True` to add a distinct marker for the arithmetic average. These coded values allow for a precise fine-tuning of the styles, box widths, and cap lengths to make the volatility in the Fresh category even more visually distinct.

Figure 10-5 shows that the Fresh category demonstrates the highest volatility, evidenced by its tall interquartile range (the height of the box) and long whiskers, indicating that fresh produce prices vary

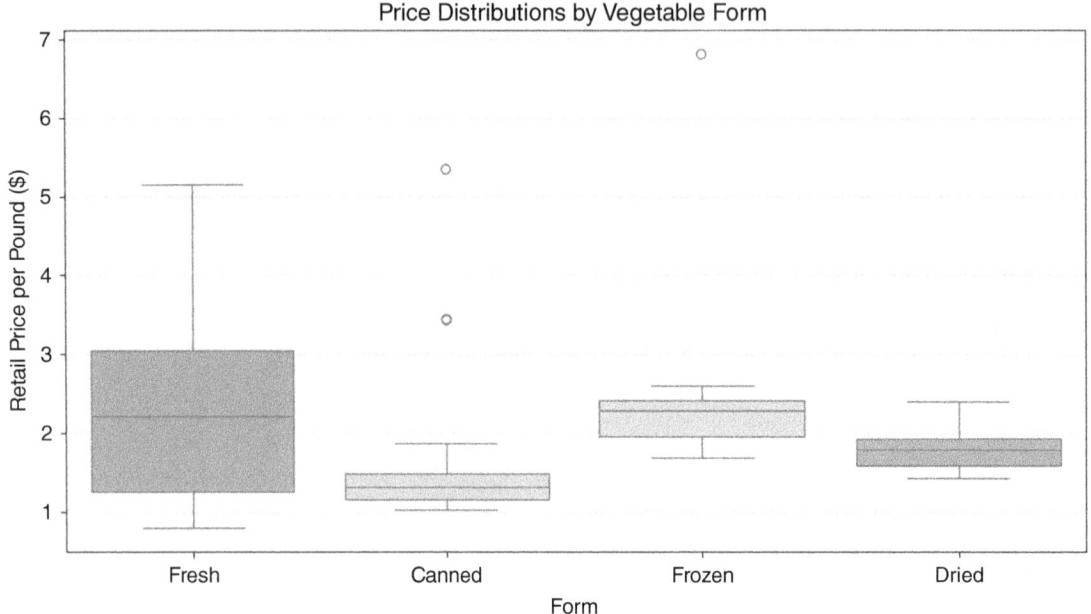

FIGURE 10-5: Boxplot.

widely from under $1 to over $5 per pound. In contrast, canned vegetables (the gray box in the middle) show a highly compressed distribution with a low median price, suggesting stability and consistency in pricing, though the distinct dots appearing above the whiskers reveal specific outliers, likely specialty items, that break this trend. The Frozen category shows a tighter price spread than fresh produce but contains the single most extreme outlier in the dataset, reaching nearly $7 per pound. Finally, the Dried category does not show outliers and has a median price around $2.

CORRELATIONS IN THE CROSS SECTION

Analyzing cross-sectional data requires more than just looking at individual distributions; it requires an investigation into how different variables interact with one another. To uncover these relationships, we primarily rely on two distinct but complementary visualization techniques. First, we use scatterplots to observe the raw, granular interaction between two or more variables, allowing us to spot nonlinear patterns and individual outliers. Second, we utilize correlation heatmaps to provide a high-level statistical summary of the entire dataset, using color-coded grids to quantify the strength of relationships between all numerical variables at once. By combining these two views, you can move from identifying broad trends to inspecting the specific data points that drive them.

Scatterplots

Cross-sectional data shines when you want to understand how two different variables interact. This is the domain of correlation. In our vegetable dataset, we have a unique variable called Yield. This represents the percentage of the vegetable that is edible (e.g., a yield of 1.0 means you eat the whole thing; 0.5 means half is waste, like peels or seeds).

We might ask: Do vegetables with higher waste (lower yield) per pound? The scatterplot is the primary tool for this investigation. It maps one variable to the x-axis and another to the y-axis. However, we can enhance this 2D plot to show four dimensions of data by utilizing Color (Hue) to represent the form (Fresh vs. Canned) and Size to represent the cost per cup (the true cost of eating).

Listing 10-6 creates a scatterplot to view the correlation. This listing builds on the vegetable prices data.

LISTING 10-6: MULTIVARIATE SCATTERPLOTS

```
import pandas as pd
import matplotlib.pyplot as plt
import seaborn as sns

# 1. Load the Data
url = "https://github.com/bkrayfield/Applied-Math-With-Python/raw/refs/heads/main/
Data/Vegetable-Prices-2022.csv"
veg_prices = pd.read_csv(url)
veg_prices = veg_prices[veg_prices['Form'].isin(['Fresh', 'Canned', 'Frozen'])]

# Visualization
plt.figure(figsize=(12, 8))

# We map 4 variables onto one chart:
# 1. x-axis: Yield (Efficiency: 0.0 to 1.0)
# 2. y-axis: RetailPrice (Cost to buy)
# 3. hue: Form (Fresh, Canned, Frozen, etc.)
# 4. size: CupEquivalentPrice (True cost to eat)
sns.scatterplot(
    data=veg_prices,
    x='Yield',
    y='RetailPrice',
    hue='Form',
    size='CupEquivalentPrice',
    sizes=(20, 200), # Control the min and max dot size for readability
    alpha=0.6,       # Transparency helps when dots overlap
    palette='viridis'
)

plt.title('Vegetable Prices: Retail Cost vs. Edible Yield', fontsize=16)
plt.xlabel('Yield (1.0 = 100% Edible)', fontsize=16)
plt.ylabel('Retail Price per Pound ($)', fontsize=16)
plt.legend(bbox_to_anchor=(1.05, 1), loc='upper left', fontsize=16) # Move
legend outside
plt.tight_layout()
plt.show()
```

Listing 10-6 demonstrates the declarative power of Seaborn, allowing us to map four distinct dimensions of data onto a single 2D plane without writing complex loops. The hue='Form' parameter instructs Seaborn to inspect the `Form` column and automatically assign a distinct color to each category (Fresh, Canned, and Frozen). Simultaneously, the `size='CupEquivalentPrice'` parameter maps the calculated serving cost to the physical area of the marker. The `sizes=(20, 200)` argument is a normalization tuple; it clamps the minimum dot size to 20 pixels and the maximum to 200 pixels, ensuring that cheap items are still visible while expensive items do not obscure the entire plot, as you can see in Figure 10-6.

We also introduce the `alpha=0.6` parameter. In scatterplots with high data density, points often stack on top of each other (called *overplotting*). By setting `alpha` (opacity) to 60%, overlapping points appear darker, revealing density clusters that would otherwise be hidden. Finally, `bbox_to_anchor` moves the legend outside the plot area, ensuring it doesn't cover our data points. The resulting chart shown in Figure 10-6 reveals if Fresh items (likely one color) cluster in a different price/yield quadrant than Canned items.

The output of the listing shown in Figure 10-6 is a multivariate scatterplot that illustrates the relationships between four different attributes of the vegetable dataset. The position of each point is determined by its yield on the x-axis (where 1.0 indicates 100% edible) and its retail price per pound on the y-axis. Furthermore, the chart uses color to distinguish between the vegetable's form—fresh, canned, or frozen—and size to represent the `CupEquivalentPrice`, where larger bubbles indicate a higher cost per edible serving. This multi-dimensional view reveals distinct clusters: frozen vegetables

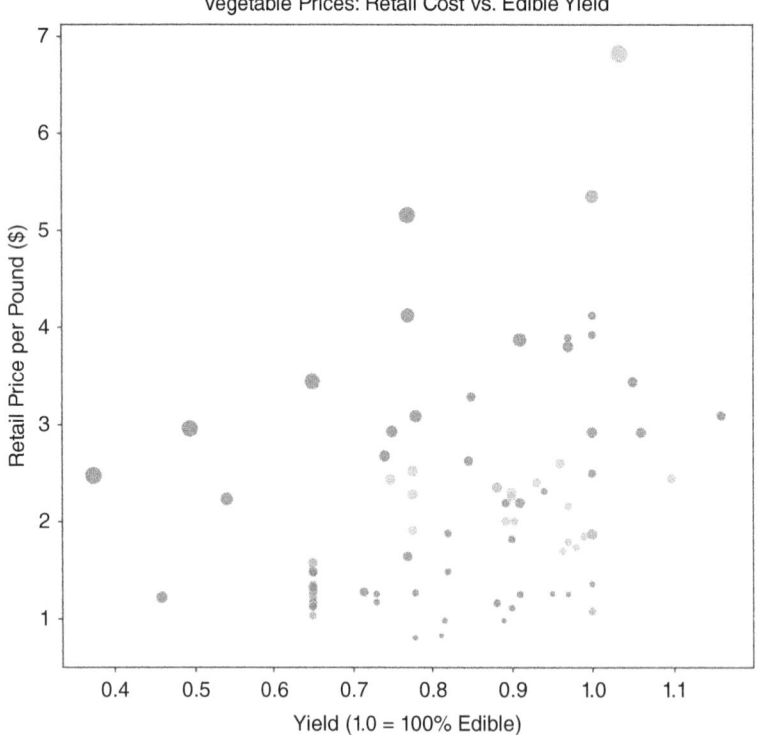

FIGURE 10-6: Multivariate scatterplot.

tend to be high-yield and lower-priced, while fresh vegetables show much greater variability across both yield and price, with several large bubbles indicating a high true cost to eat despite a moderate retail price.

Seaborn's scatterplot function provides further customization attributes to handle complex data relationships. Beyond hue and size, you can utilize the `style` parameter to map a categorical variable to the shape of the markers (e.g., squares for one category, circles for another), which is particularly useful for printing in black and white. The `markers` argument works in tandem with style to define exactly which symbols to use. For aesthetic refinement, `edgecolor` and `linewidth` allow you to add borders to your points, helping them stand out against the background. Additionally, if you are plotting a very large dataset where overplotting makes points indistinguishable, you can switch from a scatterplot to a joint plot (using `sns.jointplot`), which adds histograms or density curves to the margins of the chart to visualize the distribution of each variable independently.

Correlation Heatmaps

When you have a dataset with many numeric variables, plotting a dozen scatterplots to check for relationships is inefficient. You need a summary view, which provides a way to scan the entire dataset for connections at a single glance. For this, you can use a correlation heatmap.

A heatmap replaces numbers with colors. It visualizes the correlation matrix, a table where every variable is compared to every other variable using the Pearson correlation coefficient. This coefficient ranges from a perfect positive correlation to a perfect negative correlation, indicating no relationship.

In our vegetable data, we have `RetailPrice`, `Yield`, `CupEquivalentSize`, and `CupEquivalentPrice`. How do these metrics relate? Does a larger serving size imply a higher price? We can investigate these questions using a heatmap in Listing 10-7.

LISTING 10-7: THE CORRELATION HEATMAP

```
# 1. Prepare the Data
# Select only numeric columns for correlation calculation
numeric_cols = ['RetailPrice', 'Yield', 'CupEquivalentSize', 'CupEquivalentPrice']
correlation_matrix = veg_prices[numeric_cols].corr()

# 2. Visualization
plt.figure(figsize=(8, 6))

# Create the Heatmap
sns.heatmap(
    correlation_matrix,
    annot=True,          # Write the data value in each cell
    fmt=".2f",           # Format to 2 decimal places
    cmap='coolwarm',     # Blue (negative) to Red (positive)
    vmin=-1, vmax=1,     # Anchor the colormap range
    linewidths=0.5,      # Space between cells
    square=True          # Force cells to be square
)
```

```
plt.title('Correlation Matrix of Vegetable Metrics', fontsize=14)
plt.show()
```

Listing 10-7 begins by filtering the DataFrame to strictly numeric columns; attempting to run a correlation on text data (like vegetable names) will result in an error. We then call the .corr() method on this subset. This is a pure statistical operation that returns a new DataFrame where the indices and columns are identical and the values represent the Pearson coefficient. The resulting visualization of the heatmap is shown in Figure 10-7.

The visualization is handled by sns.heatmap. The cmap='coolwarm' argument is crucial here. It utilizes a diverging colormap, where distinct colors represent the extremes (blue/light for –1 and red/dark for +1). A neutral color (white or gray) represents the middle (0). This allows the analyst to instantly spot strong relationships. We set vmin=-1 and vmax=1 to anchor the color scale; without this, the colors would scale relative to the data (e.g., the darkest shade might only represent 0.5),

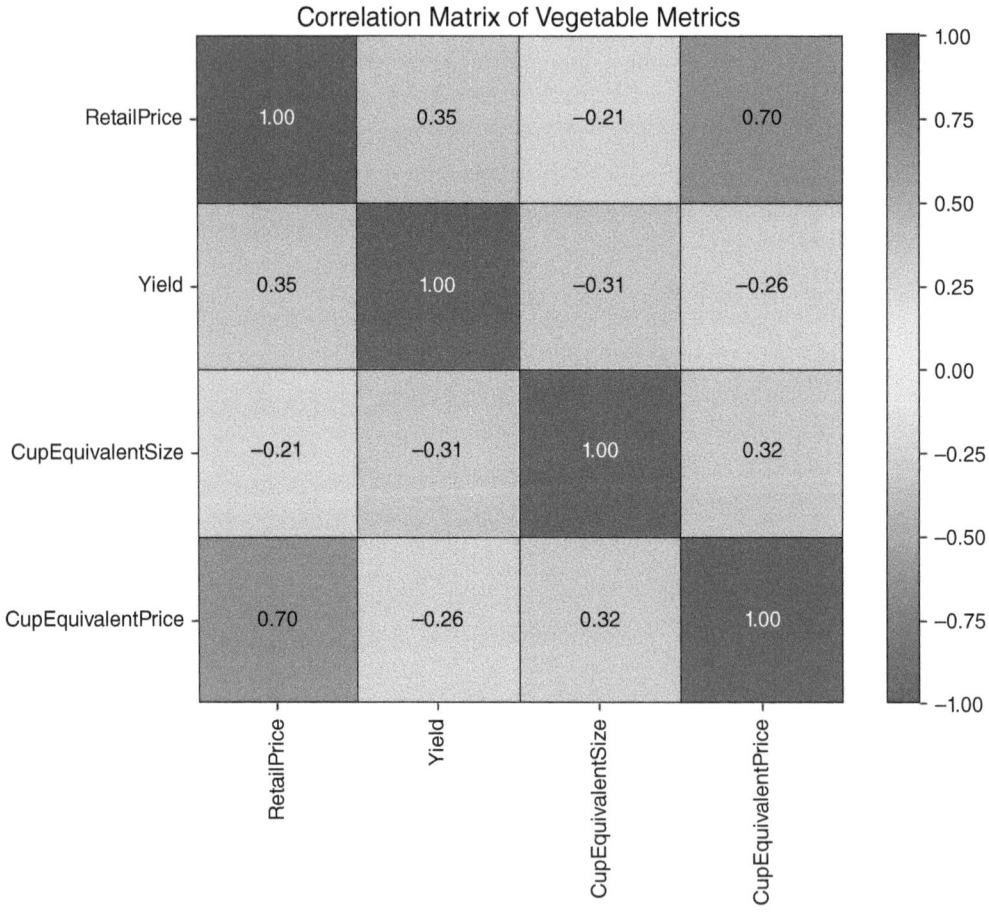

FIGURE 10-7: Stacked bar chart.

which could be misleading. Finally, `annot=True` overlays the actual correlation coefficients onto the squares, combining the visual intuition of the colors with the statistical precision of the numbers.

The result allows you to instantly see, for example, if `Yield` has a negative correlation with `RetailPrice`. The color intensity serves as a signal of the strength and direction of the relationship—dark squares indicate a positive correlation (as one number goes up, the other goes up), while light squares indicate a negative correlation. The strongest relationship, marked by the darkest square, is between `RetailPrice` and `CupEquivalentPrice`, confirming that vegetables with a higher price per pound generally translate to a higher cost per edible serving. Conversely, the light squares (such as `Yield` vs. `CupEquivalentSize`) highlight weak inverse relationships, suggesting that larger or more efficient vegetables do not necessarily correlate with higher prices.

The Pair Plot

In the previous section, we used a heatmap to find correlations. However, a single number, like a correlation coefficient, can be misleading. It tells you two things are related, but it doesn't tell you how. Is the relationship a straight line? Is it a curve? Is it driven entirely by three massive outliers?

To answer this, you need to see the raw data. Plotting every combination of variables manually (price vs. yield, price vs. size, and yield vs. size) is tedious. The pair plot automates this. It constructs a grid of charts where the diagonal shows the distribution of a single variable (histogram or kernel density estimate), and the off-diagonal cells show the relationship between two variables (scatterplot).

This visualization allows you to absorb the entire structure of your dataset in seconds.

Listing 10-8 utilizes `sns.pairplot`, one of the most powerful functions in the Seaborn library. By passing the `hue='Form'` argument, we instruct the function to not only plot the data but to segment it by category.

LISTING 10-8: GENERATING A SCATTER MATRIX WITH SEABORN

```
# 1. Prepare the Data
# We select the numeric metrics and one categorical column ('Form') for coloring
cols_to_plot = ['RetailPrice', 'Yield', 'CupEquivalentPrice', 'Form']
subset = veg_prices[cols_to_plot]

# 2. Visualization
# sns.pairplot automatically detects numeric vs. categorical data
sns.pairplot(
    subset,
    hue='Form',            # Color the dots/lines by the Vegetable Form
    palette='viridis',     # Use a distinct color scheme
    height=2.5,            # Size of each small subplot
    plot_kws={'alpha': 0.6} # Make dots slightly transparent
)

plt.suptitle('Pairwise Relationships in Vegetable Data', y=1.02, fontsize=16)
plt.show()
```

The result of Listing 10-8 is a matrix of plots shown in Figure 10-8. These plots include:

➤ **The diagonals:** Instead of scatterplots, these show the distribution of each variable. We can instantly see that RetailPrice has a long tail (a few very expensive items), while Yield is bi-modal (items are either very efficient or very wasteful, with few in between).

➤ **The scatterplots:** We can spot the relationships. For example, looking at the intersection of Yield and RetailPrice, we might see that low-yield items (like corn on the cob) tend to have lower retail prices per pound, effectively balancing out the cost to the consumer.

➤ **The clusters:** The colors reveal if certain forms behave differently. We might see that frozen vegetables cluster tightly in a specific price/yield range, while fresh vegetables are scattered all over the map.

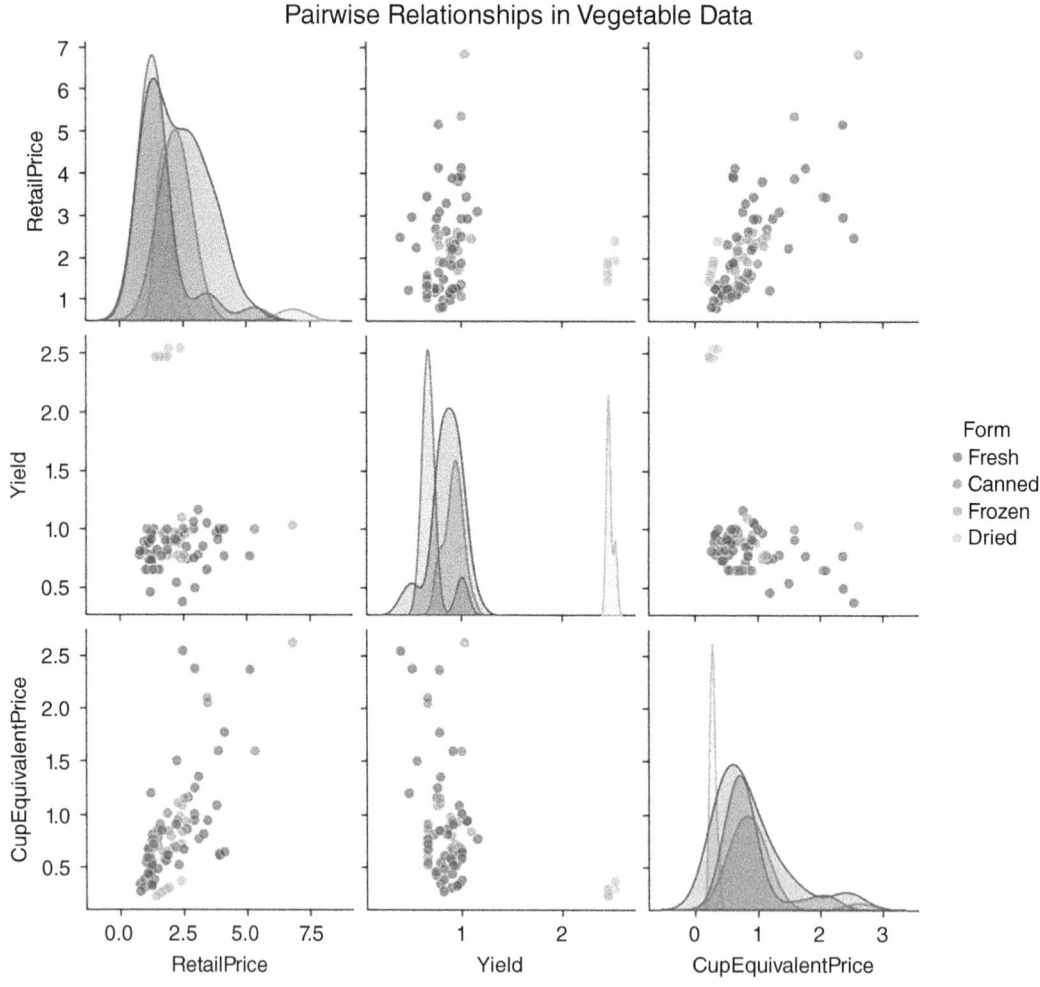

FIGURE 10-8: The pair plot.

This one command effectively replaces a dozen individual chart queries, making it the ideal starting point for any cross-sectional analysis.

Beyond basic coloring with the hue parameter, `sns.pairplot` offers extensive control over its grid through specialized keyword arguments. While the `plot_kws` parameter applies styling (like transparency or point size) to every scatterplot in the grid, you can use `diag_kws` to specifically modify the diagonal charts—for instance, by changing a kernel density estimate (KDE) to a histogram or adjusting the line thickness. For even more granularity, the `vars` parameter allows you to limit the plot to specific columns, preventing the grid from becoming overwhelming in large datasets. Furthermore, the `kind` parameter can transform the off-diagonal plots from standard scatterplots into regression plots (`kind='reg'`), which automatically adds a line of best fit to help visualize trends across vegetable forms.

Table 10-1 shows some of the customizations you can add to a pair plot.

TABLE 10-1: Pair Plot Customization Options

PARAMETER	FUNCTION	EXAMPLE
`kind`	Changes the off-diagonal plot type.	`'reg'` for regression lines; `'hist'` for 2D histograms.
`diag_kind`	Changes the diagonal plot type.	`'kde'` for smooth curves; `'hist'` for bars.
`markers`	Assigns different shapes to categories.	`markers=["o", "s", "D"]` for Fresh, Canned, Frozen.
`corner`	Removes the redundant upper triangle.	`corner=True` to create a cleaner, triangular grid.
`diag_kws`	Dictionary of properties for diagonal plots.	`{'fill': True}` to color under the KDE curve.

SUMMARY

This chapter shifted the analytical focus from the moving timeline of history to the static, high-resolution snapshot of cross-sectional analysis. It established that while time-series ask how did we get here, cross-sectional analysis indicates where we are right now. It does this by examining structure, rank, and relationship at a single distinct moment.

The chapter began by explaining the hierarchy of comparison using bar charts, noting that horizontal orientation and sorting are essential for readability when dealing with long category names or ranking tasks.

It then addressed the challenge of visualizing composition. While acknowledging the popularity of pie charts for showing part-to-whole relationships, it explored the cognitive difficulties of comparing angles versus lengths. The chapter introduced the donut chart as a clearer circular alternative and the stacked bar chart as a space-efficient, linear solution for comparing proportions.

Finally, this chapter explored techniques for revealing hidden patterns and distributions. It utilized multivariate scatterplots to map up to four dimensions of data—such as x, y, color, and size—onto a single plane. We then scaled this analysis up using correlation heatmaps to instantly scan for positive and negative relationships across all numeric variables, and pair plots to visualize the structure of those relationships. The chapter concluded by using boxplots to move beyond simple averages, allowing us to visualize volatility, spread, and outliers within our categorical groups.

CONTINUE YOUR LEARNING

Cross-sectional visualization is the most common form of business reporting. To master these charts, you must become comfortable with the specific arguments and formatting options within the Seaborn and Matplotlib libraries. The following resources and reference table will help you deepen your understanding of the tools introduced in this chapter:

➤ **Seaborn Categorical Data:** Detailed guides on bar charts, boxplots, and violin plots. `https:/ /seaborn.pydata.org/tutorial/categorical.html`

➤ **Matplotlib Pie Charts:** The official documentation for creating and customizing pie and donut charts. `https://matplotlib.org/stable/gallery/pie_and_polar_charts/pie_ features.html`

➤ **Seaborn Distribution Plots:** Learn more about pair plots and complex distribution visualizations. `https://seaborn.pydata.org/tutorial/distributions.html`

Essential Cross-sectional Functions

Table 10-2 serves as a quick reference for the essential functions used to visualize structure and relationships in this chapter.

TABLE 10-2: Key Summary Statistics Functions

STATISTIC	FUNCTION	DESCRIPTION
`.sort_values()`	Pandas	Sorts the DataFrame. Essential before plotting bar charts to create a readable staircase effect.
`sns. barplot(orient='h')`	Seaborn	Creates a bar chart. Using horizontal orientation helps read long category labels.

STATISTIC	FUNCTION	DESCRIPTION
`plt.pie()`	Matplotlib	Generates a circular composition chart. Requires a single array of values rather than X/Y coordinates.
`plt.Circle()`	Matplotlib	Used to draw a white circle over a pie chart to create a donut chart aesthetic.
`plt.barh(left=...)`	Matplotlib	Creates horizontal bars. By calculating the left parameter, we can chain bars together to create stacked bar charts.
`sns.scatterplot(hue=, size=)`	Seaborn	Plots relationships between two variables, adding color (hue) and bubble size (size) for extra dimensions.
`.corr()`	Pandas	Calculates the Pearson correlation coefficient matrix for all numeric columns in a DataFrame.
`sns.heatmap()`	Seaborn	Visualizes a correlation matrix using color intensity to show relationship strength.
`sns.pairplot()`	Seaborn	Generates a grid of scatterplots and histograms to visualize every variable against every other variable.
`sns.boxplot()`	Seaborn	Visualizes the distribution, median, and outliers of data based on the five-number summary.

11

Illustrating Alternative Data Types

In traditional econometrics and financial analysis, data is almost exclusively structured. It arrives in neat, tabular formats, rows of observations and columns of variables, ready for immediate ingestion by statistical software. However, the proliferation of digital footprints has given rise to alternative data: information derived from non-traditional sources that acts as a proxy for economic or behavioral activity.

Alternative data includes satellite imagery tracking retail parking lots, credit card transaction logs, social media sentiment, web scraping of product prices, and blockchain ledgers. The defining characteristic of this data is that it is often unstructured or semi-structured. It does not fit naturally into an X-Y plane.

The challenge of illustrating alternative data is one of abstraction. We cannot simply plot the raw data; we must first transform qualitative signals (words, locations, links) into quantitative geometry. This chapter explores the distinct visualization grammars required for text, space, and networks.

TEXTUAL ANALYSIS

The previous chapters dealt with structured data. This is information that fits neatly into rows and columns, prices, dates, quantities, and coordinates. It is numerical, sortable, and ready for calculation.

Text, however, is the largest source of unstructured data available to the modern analyst. It is messy, subjective, and highly variable. A single sentiment (e.g., "This food is bad") can be expressed in thousands of different ways ("gross," "inedible," "yuck," "not my favorite"). A spreadsheet cannot natively sum these up.

To visualize text, we must first perform tokenization. This is the process of breaking a stream of natural language into measurable units, known as *tokens* (usually individual words). Once the text is broken into tokens, we can count them, measure their sentiment, and map their relationships. We essentially transform qualitative human language into quantitative data.

Before we visualize, we must understand our source material. In this chapter, we analyze a dataset titled `Restaurant reviews.csv`.

This dataset is a classic example of alternative data. While a restaurant's financial ledger tells you how much money they made, it doesn't tell you why. The review data—specifically the unstructured text written by customers—contains the answer. It holds the *why* behind the revenue.

This dataset contains three critical columns:

➤ **Restaurant:** The entity being reviewed.

➤ **Review:** The unstructured text we want to analyze.

➤ **Time:** The timestamp, allowing us to track changes over history.

The first step is always inspection. We never start analyzing data blindly. We must load the file and print the first few rows to verify that the data was read correctly, check for missing values, and understand the column names. The following code snippet will open and show the first few rows of data:

```
import pandas as pd # Load the dataset
url = "https://github.com/bkrayfield/Applied-Math-With-Python/raw/refs/heads/
main/Data/Restaurant%20reviews.csv" df = pd.read_csv(url)
# Display the first few rows to understand the structure print(df.head())
```

Running this code reveals the tabular structure of our data. You will see a review column filled with sentences like "The chicken was dry" or "Great service!"

Crucially, you will notice the dataset contains reviews for multiple different restaurants. For our analysis, context is king. The vocabulary used to describe a good burger ("juicy," "greasy") is very different from the vocabulary used to describe good sushi ("fresh," "light"). If we analyze all restaurants together, these signals will cancel each other out.

Therefore, our strategy is to filter the data and focus our analysis on a single restaurant: Chinese Pavilion. We do this by looking at a simple word cloud that highlights the most frequent words in the customer reviews.

The Word Cloud

The most recognizable visualization in textual analysis is the *word cloud*. In this chart, the font size of a word is directly proportional to its frequency in the corpus. While it is not a tool for precise statistical comparison, it is difficult for the human eye to judge if one word is 10% larger than another, it is unrivaled for generating a high-level view of the data. It answers the immediate question: What are the dominant themes in this text?

To create a meaningful word cloud, we cannot simply feed the raw dataset into the visualization engine. We must perform three distinct data preparation steps:

1. **Filtration:** Isolate the specific entity we want to analyze to avoid context pollution.

2. **Aggregation:** Combine all individual rows of text into a single, massive string (the *corpus*).

3. **Cleaning:** Remove *stop words* (common grammatical fillers like "the," "and," "is") that would otherwise dominate the image.

Listing 11-1 implements this workflow. The dataset is filtered to isolate reviews for Chinese Pavilion, aggregate the text, and generate the cloud. Note the use of the `wordcloud` library, which handles the complex task of tokenizing the text and calculating pixel sizes automatically.

LISTING 11-1: GENERATING A WORD CLOUD

```
import pandas as pd
import matplotlib.pyplot as plt
from wordcloud import WordCloud

# 1. Load the dataset
url = "https://github.com/bkrayfield/Applied-Math-With-Python/raw/refs/heads/main/
Data/Restaurant%20reviews.csv"
df = pd.read_csv(url)

# 2. Filter for the specific restaurant "Chinese Pavilion"
# Note: Adjust column names ('Restaurant', 'Review') if they differ in your
specific CSV version
restaurant_name = "Chinese Pavilion"
subset = df[df['Restaurant'] == restaurant_name]

# 3. Combine all reviews into a single text string
# We cast to string to handle any potential missing values or non-string data
text_corpus = " ".join(subset['Review'].astype(str).tolist())

# 4. Generate the word cloud
# We use 'stopwords' to automatically remove common words like "the", "and", "is"
cloud = WordCloud(width=800, height=400, background_color='white').
generate(text_corpus)

# 5. Display the image
plt.figure(figsize=(10, 5))
plt.imshow(cloud, interpolation='bilinear')
plt.axis("off")
plt.title(f"Most Frequent Words: {restaurant_name}")
plt.show()
```

The process begins by establishing the computational environment through the importation of Pandas for data manipulation, Matplotlib for graphical rendering, and `wordcloud` for the specific text-processing algorithm. With the libraries in place, the script ingests the raw data from a remote repository using `pd.read_csv`, loading the mixed collection of reviews into a structured DataFrame. To ensure the analysis reflects a specific entity rather than a generic aggregate, the code employs Boolean indexing to filter the dataset. By selecting only rows where the "Restaurant" column matches "Chinese Pavilion," the script isolates the relevant signal from the noise of unrelated establishments.

Since the word cloud algorithm requires a single continuous block of text rather than a column of disjointed entries, the script performs a critical aggregation step. It converts the filtered review column into a list of strings and joins them together, separated by spaces, creating a unified text corpus ready for tokenization. The visualization is then constructed by instantiating a `WordCloud` object. During this generation phase, the library automatically tokenizes the corpus, removes standard stop words (such as "the" or "and"), and calculates the frequency distribution of the remaining vocabulary to determine the relative size of each term.

The code uses `interpolation='bilinear'` in the final plotting step. This is a graphical smoothing technique. Since the generated word cloud is essentially a low-resolution image, this argument blurs the rough pixel edges, making the text appear sharper and more professional for publication.

When you generate this figure, the algorithm counts every unique word in the text corpus. The words that appear most frequently are rendered in the largest font size and placed near the center. Figure 11-1 shows the words good, chicken, and food dominating the image.

When specific dishes (e.g., noodle or soup) appear large, they are the key drivers of the customer experience. However, if words like "wait," "rude," or "cold" appear in significant sizes, they act as early warning signals for operational failures.

N-grams

The primary limitation of the word cloud and simple word counts is the removal of context. The *bag of words* model treats every word as an independent entity. This concept fails to capture the relationship between adjacent words.

For example, in a restaurant review, the word "service" is neutral. It conveys the topic but not the quality. However, the sequence "quick service" is positive, while "slow service" is negative. Furthermore, there is a negation problem, where "not good" is stripped of the word "not" during stop-word removal, leaving only "good." This can clearly invert the analytical conclusion.

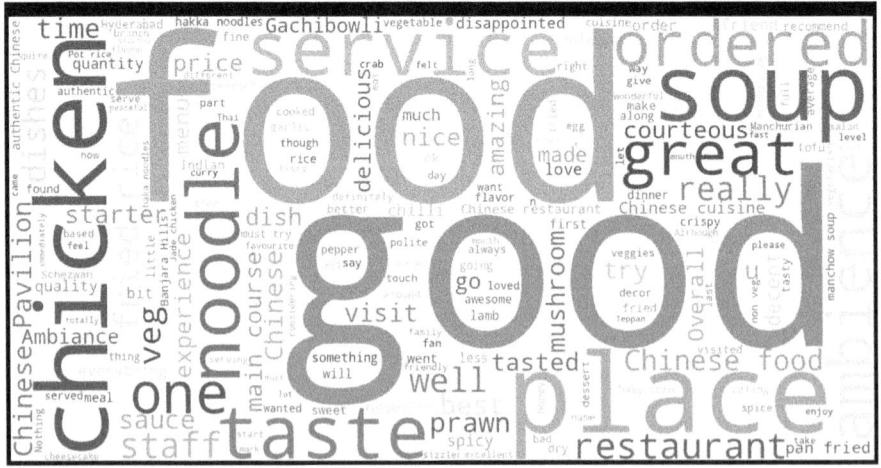

FIGURE 11-1: Word cloud.

To recover this semantic structure, we illustrate N-grams. An *N-gram* is a contiguous sequence of items from a given sample of text. Examples of various types of N-grams include:

- ➤ Unigram: Chicken

- ➤ Bigram: Spicy chicken

- ➤ Trigram: Thai spicy chicken

For most exploratory data analysis, the bigram is the sweet spot. It provides specific context (adjective + noun) without being so specific that it becomes unique to a single review (which often happens with trigrams or sentences).

To visualize bigrams, we must move beyond the standard string manipulation used for word clouds and employ a tokenizer. We utilize the `CountVectorizer` from the scikit-learn library. This tool converts a collection of text documents into a matrix of token counts, effectively turning unstructured text into a structured numerical grid.

Listing 11-2 configures the vectorizer with `ngram_range=(2, 2)`, instructing it to ignore single words and strictly identify pairs. It also applies a stop word filter to ensure our pairs are not dominated by grammatical noise, such as "in the" or "of a."

LISTING 11-2: EXTRACTING AND VISUALIZING TOP BIGRAMS

```
from sklearn.feature_extraction.text import CountVectorizer
import matplotlib.pyplot as plt
import pandas as pd

# 1. Prepare the Data
url = "https://github.com/bkrayfield/Applied-Math-With-Python/raw/refs/heads/main/
Data/Restaurant%20reviews.csv"
df = pd.read_csv(url)
df_pavilion = df[df['Restaurant'] == "Chinese Pavilion"].copy()
# We explicitly cast to string to handle any edge-case formatting issues.
text_data = df_pavilion['Review'].astype(str)

# 2. Initialize the Vectorizer
# ngram_range=(2, 2): Look ONLY for two-word sequences.
# stop_words='english': Removes pairs containing "the", "is", "a", etc.
vec = CountVectorizer(ngram_range=(2, 2), stop_words='english')

# 3. Fit and Transform
# This creates a Sparse Matrix where rows are reviews and columns are bigrams.
bag_of_words = vec.fit_transform(text_data)

# 4. Calculate Total Frequency
# We sum down the columns (axis=0) to get the total count of each bigram
across all reviews.
sum_words = bag_of_words.sum(axis=0)

# 5. Map Vocabulary to Counts
# vec.vocabulary_.items() returns the mapping of 'word' -> column_index.
words_freq = [(word, sum_words[0, idx]) for word, idx in vec.vocabulary_.items()]
```

```
# 6. Sort and Isolate Top 10
words_freq = sorted(words_freq, key=lambda x: x[1], reverse=True)
top_10_bigrams = words_freq[:10]

# 7. Convert to DataFrame for Plotting
df_bigrams = pd.DataFrame(top_10_bigrams, columns=['Bigram', 'Frequency'])

# 8. Visualizing with a Horizontal Bar Chart
plt.figure(figsize=(10, 6))
# We use a horizontal bar (barh) to ensure long bigram text is readable.
plt.barh(df_bigrams['Bigram'], df_bigrams['Frequency'], color='purple')
plt.gca().invert_yaxis() # Invert y-axis to place the highest frequency at the top

plt.title("Top 10 Most Common Phrases (Bigrams): Chinese Pavilion")
plt.xlabel("Frequency Count")
plt.grid(axis='x', linestyle='--', alpha=0.5) # Add vertical grid lines
for precision
plt.show()
```

Listing 11-2 transitions from simple string manipulation to matrix algebra. It begins by ensuring that the text data is in a uniform string format, preventing type errors during processing. The core of the analysis is the `CountVectorizer` from the scikit-learn library. This object is initialized with two specific parameters:

➤   `ngram_range=(2, 2)` forces the tokenizer to ignore single words and exclusively identify two-word sequences (bigrams).

➤   `stop_words='english'` filters out grammatical noise.

The `fit_transform` method then converts the raw text into a sparse grid where rows represent reviews and columns represent unique bigrams. By summing this matrix along the vertical axis (`axis=0`), the code calculates the global frequency of every phrase across the entire dataset.

To interpret these results, the script maps the numerical indices of the matrix back to their English counterparts using the vectorizer's vocabulary. The resulting list of tuples is sorted in descending order to isolate the top ten most frequent pairs. Finally, the visualization is rendered as a horizontal bar chart (`barh`). This orientation is specifically chosen over a vertical chart to ensure that the longer text labels of the bigrams (e.g., "Authentic Chinese") remain readable without overlapping.

The output of this code is the horizontal bar chart shown in Figure 11-2. We choose a bar chart over a word cloud here because comparative precision is paramount. We need to see exactly how much more frequent "fried rice" is compared to "pan fried."

When analyzing the results for Chinese Pavilion, you are likely to see menu items (e.g., "sour soup," "fried rice") and service descriptors (e.g., "good service," "friendly staff").

If a negative bigram, such as "long wait" or "tasted bad," appears in the top ten, it indicates a systemic failure rather than an isolated incident. If specific dishes appear frequently in this list, they are the dishes driving the restaurant's positive or negative identity.

This technique bridges the gap between qualitative reading and quantitative analysis, allowing the analyst to measure the themes of the text rather than just the vocabulary. The business benefit speaks for itself; we can now clearly see what our customers are talking about with regard to our food. And

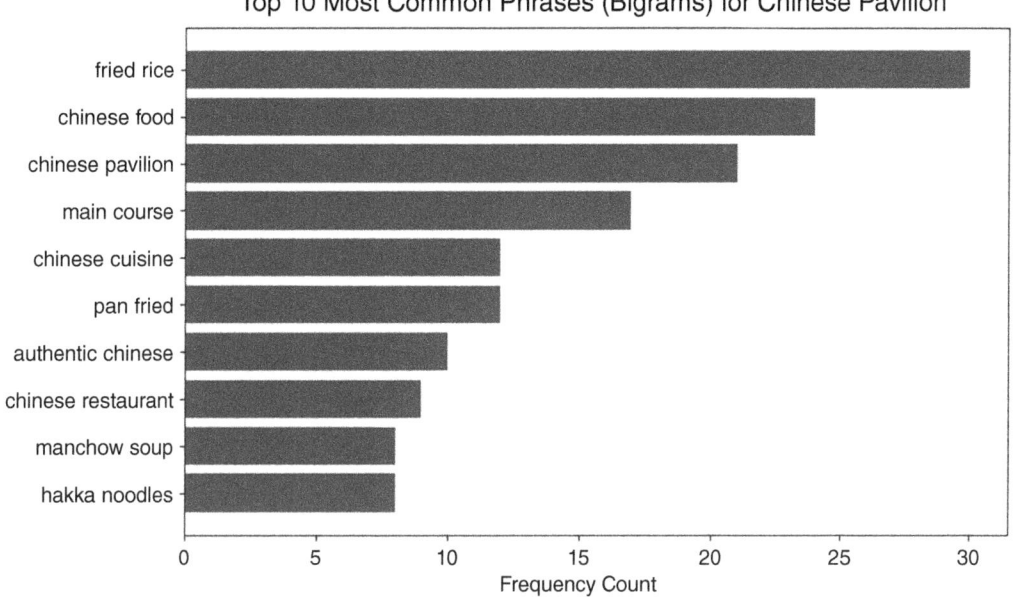

FIGURE 11-2: Bar chart with bigrams.

this is just the start of our analysis; we could do additional cleaning like removing the business name and spell checking, among others, to driver a more meaningful analysis.

Visualizing Customer Sentiment

While frequency tells us what people are saying, it doesn't tell us how they feel about it. To understand customer satisfaction, we need *sentiment analysis*. We cannot simply plot raw text; we must convert qualitative opinions into quantitative numerical scores.

To achieve this, we will use the TextBlob library, a standard tool for Natural Language Processing (NLP). Unlike complex machine learning models that require training data, TextBlob relies on a rule-based lexicon. This is essentially a predefined dictionary where thousands of adjectives and adverbs have been rated by human researchers.

Polarity is computed as follows:

➤ TextBlob assigns a polarity score to a sentence, a float value ranging from –1.0 to +1.0.

➤ Polarity > 0: Positive sentiment (e.g., "Great," "Delicious")

➤ Polarity < 0: Negative sentiment (e.g., "Terrible," "Disgusting")

➤ Polarity = 0: Neutral sentiment

The calculation is not magic; it is arithmetic. The algorithm splits the text into words and looks up each word in its lexicon. It identifies sentiment-bearing words. For example, "good" might have a score of +0.5, while "excellent" is +0.8. It looks for modifiers. If the algorithm sees "not good," it flips the polarity of "good" (multiplying by –0.5). If it sees "very good," it applies an intensifier multiplier.

Finally, it calculates the average score of all sentiment words in the text to produce the final polarity for the review. Listing 11-3 defines a function to classify each review based on this polarity score.

LISTING 11-3: PLOTTING ROLLING SENTIMENT

```python
from textblob import TextBlob

# Define a function to classify sentiment
def classify_sentiment(text):
    # TextBlob calculates polarity: < 0 is Negative, > 0 is Positive
    analysis = TextBlob(str(text))
    if analysis.sentiment.polarity > 0:
        return 'Positive'
    else:
        return 'Negative'

# Apply the function to the 'Review' column
# We create a new column 'Sentiment_Type' to store the result
df_pavilion = df[df['Restaurant'] == "Chinese Pavilion"].copy()
df_pavilion['Sentiment_Type'] = df_pavilion['Review'].apply(classify_sentiment)

# Preview the classification
print(df_pavilion[['Review', 'Sentiment_Type']].head())
```

Here, we apply the `classify_sentiment` function to every row in our DataFrame. The `.apply()` method is efficient because it works with the data like a vector. We now have a categorical variable (`Sentiment_Type`) that we can count and plot, transforming our unstructured text into structured categorical data. The result of running Listing 11-3 is a DataFrame of reviews, and a new column indicating if they are positive or negative:

```
                                          Review  Sentiment_Type
9999  Checked in here to try some delicious chinese ...        Positive
9998  I personally love and prefer Chinese Food. Had...        Positive
9997  Bad rating is mainly because of "Chicken Bone ...        Positive
9996  This place has never disappointed us.. The foo...        Positive
9995  Madhumathi Mahajan Well to start with nice cou...        Positive
```

Customer sentiment is rarely static; it fluctuates based on management changes, menu updates, and staffing issues. A static bar chart of total positive versus total negative hides this temporal story. To see the *trend*, we use a rolling window.

A *rolling window* (or *moving average*) calculates the sum of positive and negative reviews over a specific period (e.g., the last 20 reviews). This smooths out the noise of day-to-day variance and reveals the underlying trajectory of customer satisfaction. We implement this is Listing 11-4.

LISTING 11-4: PLOTTING THE ROLLING TREND

```python
import matplotlib.pyplot as plt

# Listing 11-4 should be run after Listing 11-3
```

```
# 1. Preprocess Time
df_pavilion['Time'] = pd.to_datetime(df_pavilion['Time'])
df_pavilion = df_pavilion.sort_values('Time')

# 2. Convert Sentiment to Integers (Safer than get_dummies)
# This creates the column if it's missing, or overwrites it if it exists.
# It prevents the "duplicate column" error.
df_pavilion['Positive_Count'] = (df_pavilion['Sentiment_Type'] == 'Positive').
astype(int)
df_pavilion['Negative_Count'] = (df_pavilion['Sentiment_Type'] == 'Negative').
astype(int)

# 3. Calculate Rolling Sums
window_size = 20
df_pavilion['Rolling_Pos'] = df_pavilion['Positive_Count'].rolling(window=window_
size).sum()
df_pavilion['Rolling_Neg'] = df_pavilion['Negative_Count'].rolling(window=window_
size).sum()

# 4. Plotting
plt.figure(figsize=(12, 6))

plt.plot(df_pavilion['Time'], df_pavilion['Rolling_Pos'],
         color='green', label='Positive Trend', linewidth=2)
plt.plot(df_pavilion['Time'], df_pavilion['Rolling_Neg'],
         color='red', label='Negative Trend', linewidth=2)

plt.title(f"Sentiment Trend (Rolling Window of {window_size} Reviews)")
plt.xlabel("Date")
plt.ylabel("Volume of Reviews")
plt.legend()
plt.grid(True, alpha=0.3)
plt.show()
```

Listing 11-4 begins by strictly defining the temporal order. Since rolling calculations depend on the sequence of events, the code converts the "Time" column into datetime objects and explicitly sorts the DataFrame chronologically. Without this step, the rolling window would calculate aggregates based on the arbitrary index order of the rows rather than the actual flow of time.

Next, the categorical labels are converted into quantifiable metrics. The script uses a direct Boolean mask, (df['Sentiment'] == 'Positive').astype(int) to create two parallel streams of binary data, zeros and ones, where a 1 represents the presence of a specific sentiment. This method is preferred for its precision. It's better than creating dummy variables for every possible typo or variation in the text data.

The core analysis uses the .rolling(window=20).sum() function. This technique, known as a *moving aggregate*, slides a window of 20 observations across the timeline. For each step, it sums the binary counters. This effectively smooths out the volatility of individual reviews, revealing the underlying signal, the momentum of customer opinion, rather than the noise of daily variance. Finally, the visualization plots these two smoothed trends against each other using standard Matplotlib line charts, utilizing distinct colors to allow for immediate visual comparison of the competing volumes.

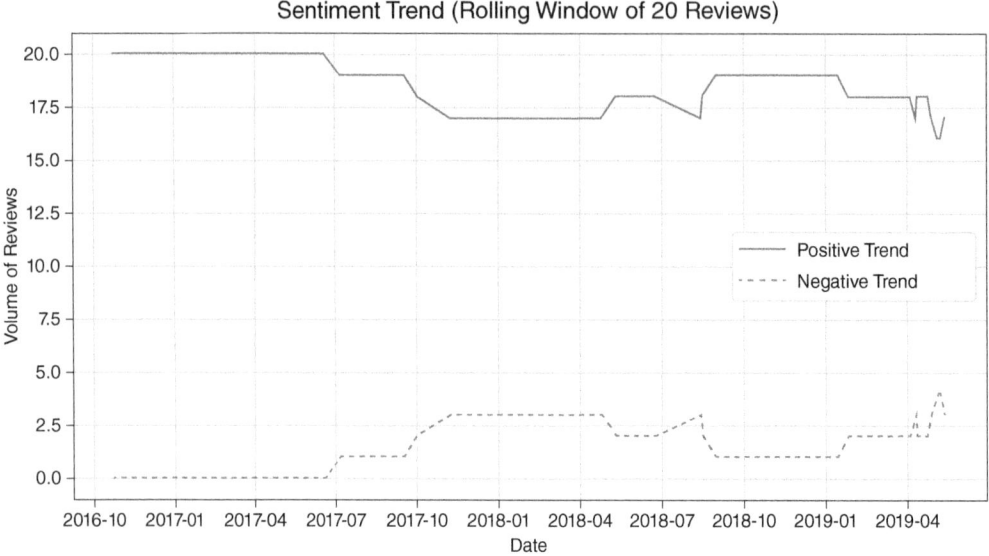

FIGURE 11-3: Sentiment over time.

The resulting line chart in Figure 11-3 plots two trends. The upper solid line represents the volume of positive feedback, and the lower dashed line represents negative feedback.

If the lower dashed line crosses above the upper solid line, it indicates a specific period where negative experiences outnumbered positive ones, a critical alert for management. If the solid line trends upward while the dashed line stays flat, the restaurant is successfully improving its reputation.

GEOSPATIAL DATA

The previous section analyzed what people were saying by examining the frequency and sentiment of their words. This section turns to examining where they are.

Geospatial data adds the dimension of physical location, specifically latitude and longitude, to our analysis. For the financial or data analyst, this is the realm of *location intelligence*. It transforms abstract rows of data into a physical reality, revealing clustering effects, regional biases, and logistics bottlenecks that are invisible in a standard spreadsheet. A table might tell you that sales are down in the Southeast, but a map instantly reveals that the drop is perfectly correlated with the path of a recent hurricane or a competitor's new distribution center.

To illustrate these concepts, we move beyond static plotting libraries like Matplotlib and utilize Folium. *Folium* is a powerful Python library that acts as a bridge to Leaflet.js, the leading open-source JavaScript library for mobile-friendly interactive maps. If you are using Google Colab or Anaconda, Folium should be included. However, if needed you can install Folium using `pip install folium`. Unlike a static image, a Folium map allows the user to zoom, pan, and click, making it the industry standard for exploring geospatial data.

The Choropleth Map

The most fundamental geospatial visualization is the *Choropleth map.* Derived from the Greek *choros* (region) and *plethos* (multitude), this technique colors geographic regions, such as countries, states, counties, or ZIP codes, based on the magnitude of a variable.

It is the standard tool for visualizing aggregate data. When you see a map of the U.S. Election results (red states vs. blue states) or a map of unemployment rates by county, you are looking at a choropleth. It is most effective when your data is already aggregated to a specific boundary level.

Creating a choropleth requires joining two distinct types of data:

➤ **Statistical data:** A standard table (CSV/DataFrame) containing the values you want to map (e.g., Population) and an identifier (e.g., State Name).

➤ **Geometric data:** A file (typically GeoJSON or TopoJSON) that defines the physical polygon boundaries of those regions.

The challenge in geospatial coding is linking these two. You must tell the code, "Match the row Alabama in my CSV to the polygon shape Alabama in the JSON file."

Listing 11-5 visualizes the population density of the United States. We load the `population_data.csv` file (which contains state names and 2020 census counts) and map it onto a standard GeoJSON definition of U.S. state boundaries.

LISTING 11-5: CREATING A U.S. POPULATION CHOROPLETH

```
import pandas as pd
import folium

# 1. Load the Statistical Data
# We read the CSV containing 'State_Name' and '2020_Pop'
df_pop = pd.read_csv(' https://raw.githubusercontent.com/bkrayfield/Applied-Math-
With-Python/refs/heads/main/Data/population_data.csv')

# 2. Define the Map Object
# We initialize the map centered on the US (Lat: 37, Lon: -95) with a
zoomed-out view
m = folium.Map(location=[37, -95], zoom_start=4)

# 3. Load Geometric Data (GeoJSON)
# We use a public URL for the US State boundaries.
# In a local project, this could be a file path like 'us-states.json'.
state_geo = 'https://raw.githubusercontent.com/python-visualization/folium/master/
examples/data/us-states.json'

# 4. Create the Choropleth Layer
folium.Choropleth(
    geo_data=state_geo,                    # The polygon shapes
    name='choropleth',
    data=df_pop,                           # The statistical data
    columns=['State_Name', '2020_Pop'],    # [Key Column, Value Column]
```

```
        key_on='feature.properties.name',    # The link: Where is the State Name in
    the JSON?
        fill_color='YlOrRd',                 # Palette: Yellow-Orange-Red
        fill_opacity=0.7,
        line_opacity=0.2,
        legend_name='Population (2020)'
    ).add_to(m)
```

The most critical parameter in this listing is `key_on`. Your CSV might have a column named `State_Name`, but the GeoJSON file might bury that information deep inside a structure like features -> properties -> name. The argument `key_on='feature.properties.name'` explicitly tells Folium how to navigate the JSON structure to find the matching string. If this key does not match your CSV column perfectly (e.g., "NY" vs. "New York"), those regions will appear gray (missing data) on the map.

We also select the `YlOrRd` (Yellow-Orange-Red) color scale. This is a sequential palette, which is best for continuous variables like population. Lighter colors intuitively suggest lower values (like Wyoming), while darker colors suggest high intensity (like California). The results can be seen in Figure 11-4.

The Marker Map

While Choropleths are excellent for aggregate regional data, they fail when we need to see specific points of interest, like individual retail stores, distribution centers, or ATM locations. For this, we use the *marker map* (see Listing 11-6).

A marker map places a pin at specific coordinates (latitude and longitude). While a simple dot is useful, Folium allows us to encode additional data into the marker using interactive aesthetics:

➤ **Popups:** Display detailed data (like revenue figures) when a user clicks the marker.

➤ **Colors:** Change the marker color based on status (e.g., Green for "Open," Red for "Closed").

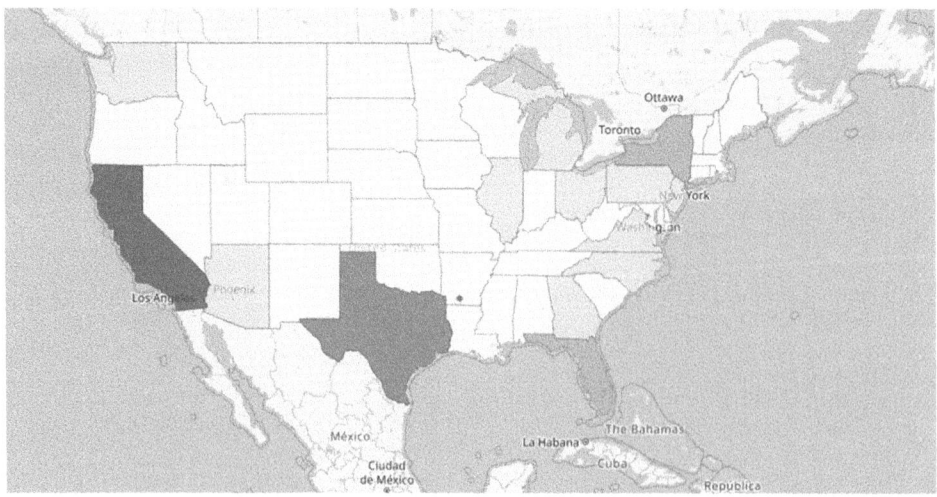

FIGURE 11-4: Choropleth map.

LISTING 11-6: CREATING AN INTERACTIVE MARKER MAP

```python
import pandas as pd
import folium

# 1. Simulate Store Data
# In practice, this would come from your internal address database
data = {
    'Store': ['Store A', 'Store B', 'Store C', 'Store D'],
    'Lat': [34.0522, 40.7128, 41.8781, 29.7604], # LA, NYC, Chicago, Houston
    'Lon': [-118.2437, -74.0060, -87.6298, -95.3698],
    'Revenue': [1.2, 3.5, 2.1, 1.8], # Millions
    'Status': ['High', 'High', 'Medium', 'Low']
}
df_stores = pd.DataFrame(data)

# 2. Initialize Map
m_markers = folium.Map(location=[37, -95], zoom_start=4)

# 3. Define a Color Helper Function
# This allows us to conditionally format markers based on data
def get_color(status):
    if status == 'High': return 'green'
    if status == 'Medium': return 'blue'
    return 'orange'

# 4. Loop Through Data and Add Markers
for i, row in df_stores.iterrows():
    folium.Marker(
        location=[row['Lat'], row['Lon']],
        popup=f"{row['Store']}: ${row['Revenue']}M", # Click to see Revenue
        tooltip=row['Store'],                         # Hover to see Name
        icon=folium.Icon(color=get_color(row['Status']), icon="info-sign")
    ).add_to(m_markers)

m_markers

# You can uncomment this line to create the map outside of a notebook environment.
# m_markers.save("my_map.html")
```

Listing 11-6 visualizes discrete entities using point data. It begins by creating a Pandas DataFrame, `df_stores`, which simulates a proprietary internal database containing store names, exact GPS coordinates (latitude/longitude), revenue figures, and a performance status.

To visualize this data, the code initializes a `folium.Map` centered on the continental United States. A critical component of this script is the helper function `get_color`. This function acts as a conditional formatter, translating the categorical `Status` variable into specific marker colors. When plotting with color, we utilize a palette of green, blue, and orange. This selection is intentional; unlike the traditional red/green/yellow traffic-light scheme, this palette is distinguishable by individuals with red-green color blindness, ensuring the visualization remains accessible to a wider audience.

The visualization is constructed via a `for` loop that iterates through every row of the DataFrame. Inside the loop, `folium.Marker` is called for each store. This method accepts arguments for location,

FIGURE 11-5: Marker map.

interactivity (popup and tooltip), and aesthetics. The icon parameter utilizes the previously defined color function, dynamically styling each pin based on its data. Finally, the code object m_markers renders the interactive map within a notebook environment. For users working in standard Python scripts (outside of Jupyter), the commented line m_markers.save("my_map.html") demonstrates how to export the visualization as a standalone HTML file viewable in any web browser.

The resulting map, shown in Figure 11-5, acts as a visual dashboard.

A regional manager can instantly scan the country and a marker on Houston. By clicking it, they access the specific revenue data. This combines the high-level overview of a chart with the granular detail of a table.

The Heatmap

There is a limit to the marker map. When the dataset grows from 50 stores to 50,000 user check-ins, markers inevitably overlap. The map becomes a chaotic "hairball" of pins, and the underlying patterns are obscured by the clutter.

To solve this, we use a *geospatial heatmap*. This visualization discards individual identity in favor of density. It applies a kernel density estimate (KDE) similar to the smoothing we use in histograms over the latitude/longitude grid.

The algorithm looks at every point and adds a small Gaussian distribution (bell curve) of intensity around it. Where points cluster together, these intensities stack up, turning the map red. This is the standard tool for analyzing high-volume alternative data like foot traffic, Uber pickup locations, or crime statistics. Running Listing 11-7 in a new cell will generate the heatmap.

LISTING 11-7: GENERATING A DENSITY HEATMAP

```
import numpy as np
from folium.plugins import HeatMap
import folium

# 1. Simulate High-Density Data
# We generate 1,000 random points clustered around NYC coordinates
lat_center, lon_center = 40.7128, -74.0060
# np.random.normal creates a "blob" of points around the center
lats = np.random.normal(lat_center, 0.05, 1000)
lons = np.random.normal(lon_center, 0.05, 1000)

# HeatMap expects a list of [lat, lon] pairs
heat_data = list(zip(lats, lons))

# 2. Initialize Map
m_heat = folium.Map(location=[lat_center, lon_center], zoom_start=10)

# 3. Add HeatMap Layer
# radius: Controls how wide the "glow" of each point is
# blur: Controls how smooth the transition between colors is
HeatMap(heat_data, radius=15, blur=20).add_to(m_heat)

m_heat
```

Listing 11-7 demonstrates how to visualize high-density geospatial data without relying on external files. It begins by utilizing NumPy to simulate a large dataset of 1,000 check-ins. The np.random. normal function generates data points that follow a Gaussian distribution around a central coordinate (New York City). This creates a realistic "cluster" effect, mimicking how human activity naturally concentrates around city centers rather than being uniformly distributed.

Data preparation is a critical step in Folium. The library's HeatMap plugin does not accept a Pandas DataFrame directly; it requires a list of coordinate pairs (e.g., [[lat, lon], [lat, lon]]). The code achieves this efficient transformation using Python's built-in zip function, which pairs the latitude and longitude arrays into a single list of tuples.

Finally, the visualization is rendered. The HeatMap layer is added to the base map with two controlling parameters. The first, radius, determines the pixel size of each point. Larger radii cause points to overlap more easily, increasing the intensity and blur, which controls the smoothness of the color gradient. A higher blur value creates a continuous, weather-radar aesthetic, while a lower value makes the individual points more distinct.

The resulting visualization, shown in Figure 11-6, looks like a weather radar. Dark zones indicate high intensity (e.g., a busy downtown district). Light zones indicate sparse activity.

By switching from markers to heatmaps, we change our analytical question. We are no longer asking where this specific entity is, but rather, where the market demand is located. This is crucial for site selection—identifying hot zones where demand exists but your physical presence (markers) does not.

FIGURE 11-6: Density heatmap.

VISUALIZING NETWORKS

The previous sections analyzed individual entities, observing what they said (text analysis) and where they were located (geospatial analysis). In standard econometrics, these entities are often treated as independent observations (rows in a spreadsheet). However, in the world of alternative data, the primary unit of analysis is frequently not the entity itself, but the relationship between entities.

Consider the following datasets:

➤ **Blockchain:** A wallet is private, but we can visualize who a wallet interacts with.

➤ **Social media:** An influencer is defined by the structure of their follower graph.

➤ **Supply chain:** A factory's risk profile is defined by its connections to upstream suppliers and downstream distributors.

This type of data cannot be visualized with cartesian (X-Y) plots. It requires network analysis (also known as graph theory). Instead of axes and coordinates, we visualize nodes (vertices) and edges (links).

To perform this analysis in Python, we utilize NetworkX. This is the industry-standard library for the creation, manipulation, and study of the structure, dynamics, and functions of complex networks.

While NetworkX handles the heavy mathematical lifting (calculating paths, centrality, and clusters), it relies on Matplotlib to render the actual visual output. The workflow typically involves defining the graph structure in NetworkX, calculating a layout (where the nodes should sit), and then passing those coordinates to Matplotlib for drawing.

Visualizing Structure

The most common challenge in visualizing networks is simple. When you have hundreds of connected nodes, how do you arrange them onscreen so the structure is readable?

The solution is the *force-directed graph* (specifically the Fruchterman-Reingold algorithm). This algorithm simulates a physical system:

➤ Nodes act like electrically charged particles that repel each other. This prevents nodes from overlapping and creates space.

➤ Edges act like elastic springs that pull connected nodes together.

When the simulation runs, the system seeks a state of equilibrium. The result is a layout where highly connected "communities" naturally cluster together, while isolated outliers drift to the periphery. This allows the analyst to instantly spot *hubs* (central influencers) and *bridges* (critical connectors between groups) without any manual sorting.

In Listing 11-8, a supply chain network is simulated to identify structural vulnerabilities. We manually define the nodes and edges, compute the force-directed layout, and render the graph.

LISTING 11-8: VISUALIZING A SUPPLY CHAIN NETWORK

```python
import networkx as nx
import matplotlib.pyplot as plt

# 1. Initialize the Graph Object
G = nx.Graph()

# 2. Define Nodes
nodes = ["Supplier A", "Manufacturer", "Distributor X", "Distributor Y",
         "Retailer Z", "Retailer W"]
G.add_nodes_from(nodes)

# 3. Define Edges
edges = [
    ("Supplier A", "Manufacturer"),
    ("Manufacturer", "Distributor X"),
    ("Manufacturer", "Distributor Y"),
    ("Distributor X", "Retailer Z"),
    ("Distributor Y", "Retailer Z"),
    ("Distributor Y", "Retailer W")
]
G.add_edges_from(edges)

# 4. Compute Layout
pos = nx.spring_layout(G, seed=42)
```

```
# 5. Draw the Graph
plt.figure(figsize=(10, 6))

nx.draw_networkx_nodes(G, pos, node_size=4000, node_color='lightblue')

# Draw the edges
nx.draw_networkx_edges(G, pos, edge_color='gray', width=2)

# Add labels
nx.draw_networkx_labels(G, pos, font_size=10, font_family="sans-serif")

plt.title("Supply Chain Network Structure")
plt.axis('off')
plt.show()
```

Listing 11-8 begins by initializing an empty graph object G using `nx.Graph()`. A graph object acts as the central data structure that stores the topological definition of the network, maintaining a registry of all nodes and the specific links that connect them, independent of how they might eventually be drawn on a screen. We then manually populate this container by defining a list of nodes (the entities) and a list of edges (tuples representing the connections between them).

The critical step occurs at `nx.spring_layout(G)`. This function executes the Fruchterman-Reingold force-directed algorithm discussed previously, calculating the coordinates for every node based on simulated physical forces. We explicitly set a seed value to ensure that the "random" initial positions are identical every time the code is run, guaranteeing a reproducible figure rather than a layout that shifts with every execution.

The visualization is then constructed in layers using Matplotlib. First, `draw_networkx_nodes` renders the entities as large, light-blue circles (sized at 4,000 to ensure visibility). Next, `draw_networkx_edges` draws the connections as gray lines, and `draw_networkx_labels` overlays the text names. Finally, `plt.axis('off')` removes the standard Cartesian grid and tick marks, leaving only the abstract network structure visible.

The network graph is generated from Listing 11-8 and shown in Figure 11-7. When you view the output, notice the position of Retailer Z versus Retailer W.

Retailer Z is pulled toward the center because it is connected to two distributors (X and Y). The springs from both sides hold it in place, visually representing supply chain redundancy.

Retailer W, however, dangles at the edge. It has only one connection. This visual simplicity highlights a complex risk: if Distributor Y fails, Retailer W is cut off, whereas Retailer Z survives.

Network analysis is not limited to supply chains; it is a universal framework for analyzing any system where connections matter more than individual attributes. In business, this "relational" perspective unlocks insights that standard tabular analysis misses, including:

> **Financial fraud detection:** In anti-money laundering (AML), criminals often use *smurfing*, breaking large sums into small, inconspicuous transfers to evade detection. A standard histogram of transaction sizes would miss this. However, a network graph reveals the structure: A single source wallet dispersing funds to hundreds of intermediary wallets, which then reconverge at a single destination. This fan-out, fan-in pattern is instantly visible in a force-directed graph.

➤ **Organizational behavior:** An org chart tells you who should have power, but a network graph of email or Slack traffic tells you who actually has influence. By mapping communication flows, HR can identify informal leaders, people who bridge disconnected departments. If these critical connectors leave the company, the organization risks becoming siloed. This metric is often called *betweenness centrality*.

➤ **Marketing attribution:** In digital marketing, a customer rarely buys a product after clicking a single ad. They might see a tweet, click a Google ad, read a blog post, and then buy. Network analysis can model the customer journey as a path through a graph, where nodes are touchpoints (Facebook, email, website). By visualizing these paths, marketers can identify the most common bridges that lead to conversion, rather than just crediting the last click.

➤ **Systemic risk in portfolio management:** Instead of viewing stocks as independent assets, network analysis views them as a correlated web. If stock A crashes, does it pull stock B down with it? By building a graph where edges represent high correlation, risk managers can visualize the centrality of a specific asset. If a portfolio is heavily invested in a hub asset, the risk of contagion is far higher than standard variance models might predict.

By applying graph theory to these domains, analysts move from asking how much to asking how is this connected?

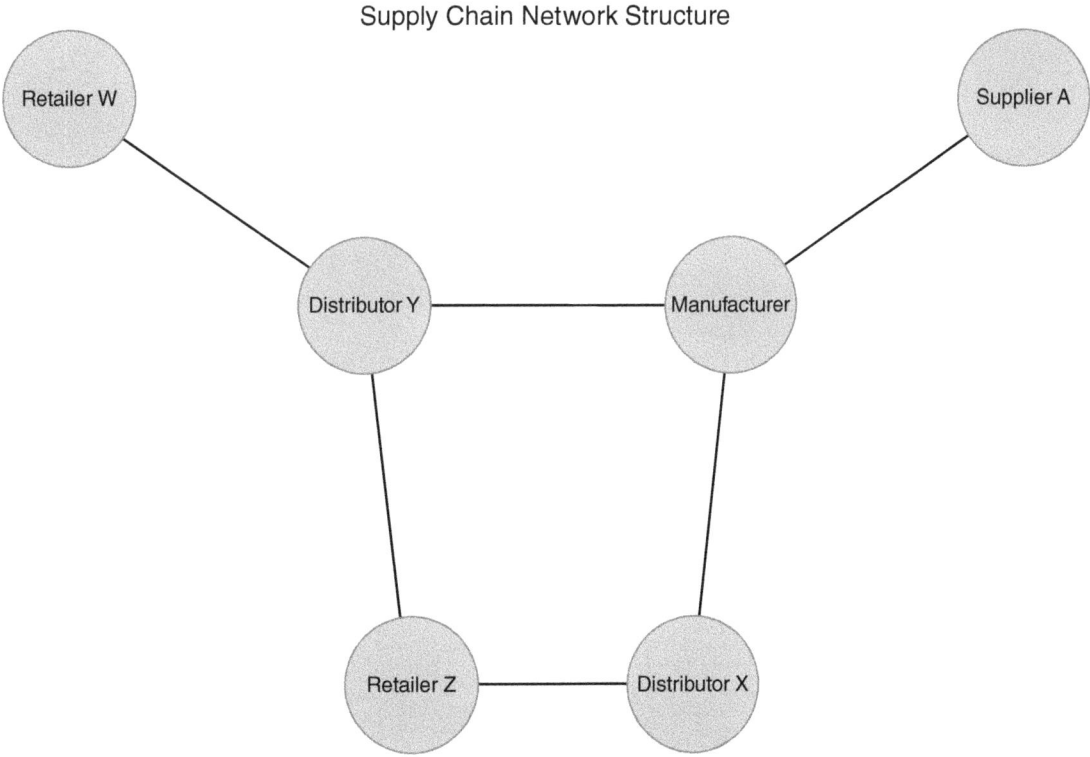

FIGURE 11-7: Simple graph.

Weighted Graphs

Real-world networks are rarely binary (connected vs. not connected). They usually possess *weights*, a measure of the strength or capacity of the connection.

➤ In social media, Close Friend is greater than (>) Acquaintance.

➤ In finance, $1M Transaction is greater than (>) $10 Transaction.

To visualize this, we map the edge width to the transaction volume. This transforms the graph from a structural map into a flow map, allowing us to see not just where connections exist, but how much activity is moving through them.

Listing 11-9 enhances our supply chain model. We assign a numerical weight to each edge representing shipment volume and use that value to control the line thickness in the visualization.

LISTING 11-9: VISUALIZING WEIGHTED FLOWS

```python
import networkx as nx
import matplotlib.pyplot as plt

# 1. Define Weighted Edges
weighted_edges = [
    ("Supplier A", "Manufacturer", 5),
    ("Manufacturer", "Distributor X", 2),
    ("Manufacturer", "Distributor Y", 3),
    ("Distributor X", "Retailer Z", 2),
    ("Distributor Y", "Retailer Z", 1),
    ("Distributor Y", "Retailer W", 2)
]

# 2. Create the Graph
G_weighted = nx.Graph()
for u, v, w in weighted_edges:
    G_weighted.add_edge(u, v, weight=w)

# 3. Compute Layout
pos = nx.spring_layout(G_weighted, seed=42)

# 4. Extract Weights for Visualization
widths = [G_weighted[u][v]['weight'] * 2 for u, v in G_weighted.edges()]

# 5. Draw
plt.figure(figsize=(10, 6))

nx.draw(G_weighted, pos,
        with_labels=True,
        node_color='lightgreen',
        node_size=6000,
        width=widths,        # Thickness corresponds to volume
        edge_color='green')
```

```
plt.title("Weighted Network: Thickness = Volume")
plt.axis('off')
plt.show()
```

Listing 11-9 focuses on translating numerical attributes into visual properties. After defining the data in Step 1, it creates the network structure (Steps 2 and 3) using a list of tuples, where each entry contains a source node, a target node, and an integer representing the weight (e.g., shipment volume). A NetworkX graph object is instantiated, and the edges are added via a loop that explicitly assigns this weight as an attribute to every connection.

To visualize the varying intensities of these relationships, the code performs a critical mapping step before plotting. The list comprehension in Step 4 generates the `widths` variable by iterating through the graph's edges, extracting the stored weight value, and multiplying it by a scaling factor of 2. This creates a vector of line thicknesses that is directly proportional to the volume of the flow.

Finally, Figure 11-8 is rendered using `nx.draw`. The calculated widths list is passed to the `width` parameter, dynamically adjusting the stroke size of each link. Simultaneously, the `node_size` is set to 6,000 to ensure the entities are prominent, resulting in a visualization where the viewer's eye is immediately drawn to the high-volume areas of the supply chain rather than the low-volume areas.

The image in Figure 11-8 tells a different story than the image in Figure 11-7. While the structure is the same, the thick line connecting Supplier A to Manufacturer immediately draws the eye, identifying

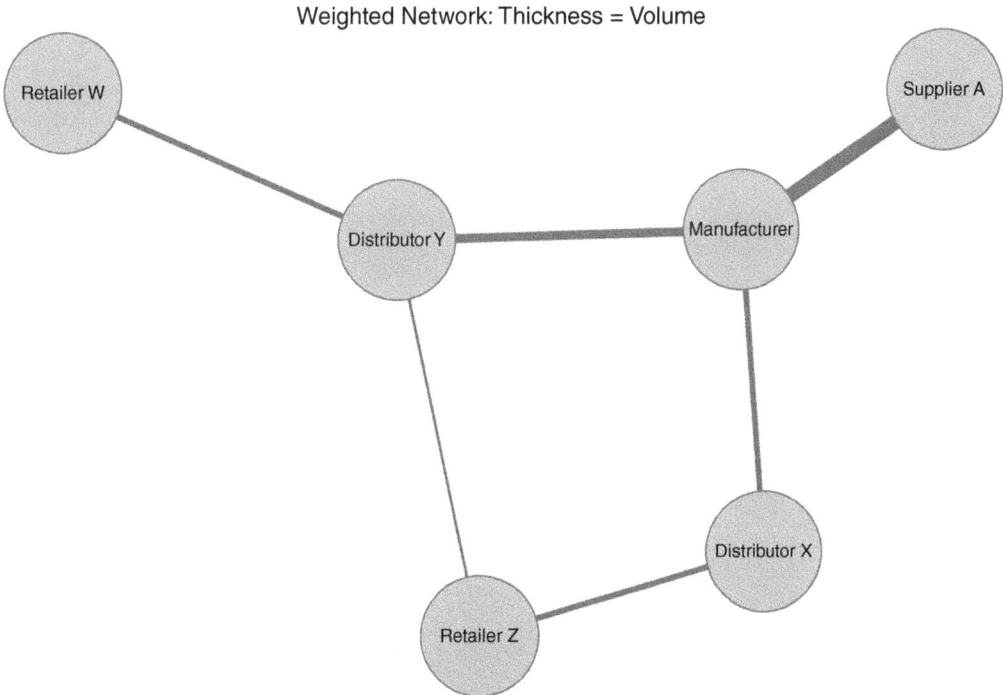

FIGURE 11-8: Weighted graph.

the critical path of the network. Conversely, the thin line between Distributor Y and Retailer Z suggests that while a connection exists, it is operationally insignificant. This technique is critical for bottleneck analysis, identifying pipes that are too small for the volume they are expected to carry.

SUMMARY

This chapter stepped beyond the rigid rows and columns of traditional structured data to explore the chaotic but rich world of alternative data. We defined this data not by its format, but by its source—unstructured digital footprints like text, location pings, and relationship webs that serve as proxies for real-world behavior.

The chapter began by tackling textual analysis, the process of quantifying human language. You learned that while word clouds provide an immediate visual summary of dominant keywords, they often strip away context. To recover this meaning, we advanced to N-gram analysis (specifically bigrams) to capture phrases and sentiment analysis using TextBlob to track the emotional trajectory of customer feedback over time. It is worth noting that these foundational concepts of sequencing words—predicting what comes next based on what came before—are the conceptual ancestors of modern Large Language Models (LLMs), which rely on massive, sophisticated embeddings of such patterns to understand and generate human-like text.

Next, the chapter moved to geospatial analysis, transforming abstract coordinates into location intelligence. We utilized the Folium library to create three distinct map types: choropleth maps for regional aggregates, marker maps for specific points of interest, and density heatmaps to visualize high-volume activity where individual points overlap.

Finally, the chapter explored network analysis using NetworkX. We shifted our focus from analyzing entities to analyzing relationships. By employing force-directed graphs, we visualized the invisible structure of supply chains, identifying central hubs, critical bridges, and vulnerabilities through both simple connections and weighted flows.

Collectively, these techniques demonstrate that the challenge of illustrating alternative data is one of translation: converting qualitative signals, words, places, and links into, geometry.

CONTINUE YOUR LEARNING

Alternative data visualization is the frontier of modern analytics. To master these unstructured formats, you must become comfortable with the specialized libraries designed for text, geography, and networks. The following resources and links to official documentation will help you deepen your understanding of the tools introduced in this chapter:

➤ **TextBlob documentation:** A simplified guide for performing common NLP tasks like sentiment analysis and part-of-speech tagging. `https://textblob.readthedocs.io/en/dev/`

➤ **Folium maps:** The official documentation for creating interactive Leaflet maps in Python, including markers and choropleths. `https://python-visualization.github.io/folium/`

➤ **WordCloud for Python:** A detailed reference for generating word clouds, including masking and coloring options. `https://amueller.github.io/word_cloud/`

➤ **NetworkX tutorial:** A comprehensive guide to creating, manipulating, and studying the structure of complex networks. `https://networkx.org/documentation/stable/tutorial.html`

INDEX

Note: Page numbers in *italics* and **bold** refers to figures and tables respectively.

A

Anaconda Notebooks, 6
apply() method, 240
arrays and matrices
  one-dimensional arrays, 18–19
  two-dimensional arrays, 19–23

B

betweenness centrality, 251
business problems with math
  constrained optimization, 173–177
  dynamic loan amortization engine, 168–171, *170*
  logistic regression, 179–185
  quality control, 177–179
  recommender system, 171–173

C

calculate_profit function, 97
calculate_similarity() function, 173
calculus
  differential equations, 93–95
  ecosystem in Python, 90–93
  marginal cost, 99–100, *100*
  marginal profit, 102–103, *103*
  marginal revenue, 100–101, *102*
  numerical differentiation and integration, 84–90
  sensitivity analysis, 96–98
classify_sentiment function, 240
code cell, 8
compound interest, 93
constrained optimum, 121

constrained *vs.* unconstrained optimization, 119–121, *121*
convex optimization, 108
cost function, 91–92
cross-sectional data
  bar charts, 218–220, *220*
  boxplots, 220–222, *222*
  correlation heatmaps, 225–227, *226*
  donut charts, 213–216, *215*
  functions, **230–231**
  pair plot, 227–229, **229**
  pie chart, 211–213, *213*
  scatterplots, 222–225, *224*
  stacked bar charts, 216–217, *218*
cumsum() function, 87
customer churn, 156–161

D

daily_new_subs array, 89
DataFrame, 23–24
data manipulation with Pandas
  columns and rows, 24–25
  creating new columns, 25–26
  DataFrame construction, 23–24
  describe() method, 24
  filtering with Booleans, 25
  grouping and aggregation, 26, *26*
  head() method, 24, *24*
  info() method, 24
  joins and merges, 27
  melt function, 28
  pivot function, 28
  stack function, 28

data types
  converting types, 17
  customer lifetime value (CLV), 16–17
  float, 15
  integer (int), 14
  margin, 15
  missing/null values, 16
  presence/absence signals, 17
  string (str), 15–16
  true/false logic, 16
describe() method, 24

E

eigenvalues and eigenvectors
  long-term customer loyalty, 77–78
  representation, 76–77
  Stock Portfolio, 78–79

F

forecasting, 161–164

G

geospatial data
  choropleth map, 243–244, 245
  heatmap, 246–247, 248
  location intelligence, 242
  marker map, 244–246, 246
Google Colab notebook, 6, 7
grouped bar chart, 44, 45–46, 46

H

head() method, 24, 24
hypothesis testing
  A/B test, 147–148
  alternative hypothesis, 145
  confidence intervals, 148–149
  null hypothesis, 145
  p-value, 146
  test statistics, 145–146

t-value (t-statistic), 145
z-value (z-statistic), 145

I

info() method, 24
integrate.quad() function, 92
interpolate() function, 198
interquartile range (IQR), 221

J

Jupyter Notebook, 6, 8

K

Kernel Density Estimate (KDE) line, 54

L

linear algebra
  matrix, 60–62
  vectors, 59, 60
linear interpolation, 197
linear programming (LP)
  constrained optimization, 113, 116
  geometry of optimization, 116–119
  infeasibility, 119
  objective function, 114, 115
  scipy.optimize.linprog function, 113, 114
  unboundedness, 119
linear regression
  extrapolation, 156
  financial risk factors, 153–155
  imply causation, 155
  marketing effectiveness, 151–153
  multicollinearity, 155
  multivariate linear regression, 149
  ordinary least squares (OLS), 150
  regression analysis, 149
  time variable, 150
linprog() function, 176
loan payoff trajectory, 168–171, 170
logistic regression, 156–161, 160

M

Markdown cell, 8
mathematical operations
  arithmetic, 13–14
  arrays and matrices, 18–23
  data manipulation, 23–28
  data types, 11, 14–17
  math module, 14
  variables, 11, 12
Matplotlib, 7, 22, *23*
`melt` function, 28
`model.predict ()` function, 178
Monte Carlo simulation, 22–23

N

NetworkX, 248, 249
numerical differentiation and integration
  inflection point, 86–87
  integration, 87–90, *89*
  mobile game downloads, 84–86, *86*
NumPy, 7
  algebra functions, **81**
  asset returns from prices, 69–70, *70*
  comparing strategies, 75–76
  constant weights, 68, 70–72, *72*
  manipulation tools, **80**
  `np.cov` function, 78
  numerical calculus, 90–91
  time-varying weights, 68, 72–74, *74*
`nx.Graph()` function, 250

O

`objective_function`, 112
optimization techniques
  constrained *vs.* unconstrained optimization,
    119–121
  constraints, 107
  CVXPY (convex optimization), 108
  four-step formulation process, 109
  linear programming (LP), 112–119
  local *vs.* global optima issue, 110, *118*
  objective function, 107
  profit maximization, 110–112
  PuLP and Pyomo, 108
  real-world applications, 122–134
  `scipy.optimize` function, 108

P

Pandas, 7
  merging methods, **27**
panel data, 190, 193–194, 206–208
`pivot` function, 28
Plotly, 7
polynomial/spline interpolation, 197
polytope, 119
portfolio allocation
  bounds and initial guess, 124
  efficient frontier, 126–127, *128*
  interpret the result, 125–126
  mock data, 122–123
  objective and constraints, 123–124
  run the optimizer, 124–125
Principal Component Analysis (PCA), 78, 79
`print` statements, 75
Python
  Anaconda Distribution, 5
  for business, 3–4
  cloud-friendly alternatives, 6
  core operators, 13, **13**
  ecosystem, 7–8
  Jupyter Notebook, 6
  libraries locally, 8–9
  spreadsheets, 4
  writing script, 9

R

random variables and distributions
  binomial distribution, 141–142, *142*
  discrete *vs.* continuous distributions, 139
  distribution of daily sales, 139
  normal distribution, 140, *141*
  Poisson distribution, 143
  uniform distribution, 142–143, *144*

real-world applications
  portfolio allocation, 122–128
  supply chain and operations, 128–131
  workforce scheduling, 131–134
rolling average, 47–49, *49*
rolling.sum() function, 241

S

scikit-learn, 7
scipy.integrate module, 92
SciPy, numerical methods, 92–93
scipy.optimize.linprog function, 113, 114
scipy.optimize.minimize_scalar
    function, 110
Seaborn, 7
seaborn's relplot function, 207
seasonal_decompose function, 205
seasonality and autocorrelation
  autocorrelation function (ACF), 201, *203*
  multiperiodicity, 202
  partial autocorrelation (PACF), 202, *203*
  Personal Consumption Expenditures (PCE), 202
  time-series decomposition, 204–205, *206*
sns.barplot() function, 219
sns.heatmap function, 226
solve_ivp function, 93, 95
stacked bar chart, 44, 45, *45*
stack function, 28
statistics ecosystems, 137–138
statistics functions, **165–166**
stats.ttest_1samp function, 178
stock_prices matrix, 68, 69
supply chain and operations
  mock data, 128–129
  objective and constraints, 129–130
  optimizing, 130–131
SymPy, symbolic calculus, 91–92

T

textual analysis
  dataset, 234
  rolling window, 240–241

  sentiment analysis, 239–242, *242*
  word cloud, 234–236
tidy data, 51
time interpolation, 197
time-series and linear data
  Consumer Price Index (CPI), 190, 192
  cross-sectional data, 189–191
  functions, **210**
  panel data, 190, 193–194, 206–208
  time-series data, 190, 192–193
  visualizing change over time, 194–206
  visualizing panel data, 190
time-series diagnostics
  aggregating and resampling, 198–200
  boxplots and histograms, 200–201
  detecting missing values, 196–197
  handling missing values, 197–198, *199*
total_profit function, 110
two-dimensional arrays
  aggregations by axis, 19–20
  broadcasting a vector, 21
  matrix multiplication, 20
  Monte Carlo simulation, 22–23
  random vectors, 22–23
  slicing and Boolean masks, 21

U

unconstrained nonlinear optimization, 110
unconstrained optimum, 121

V

vectors and matrices
  combining matrices, 64–65
  dot product, 63
  matrix multiplication, 66
  norms (vector lengths), 64
  with NumPy, 67–76
  scalar multiplication, 63
  slicing matrices, 65–66
  transpose, 67
  working with, *59–62*
viral marketing, 93

visualization
  business data communication, 37–42
  categories and segments, 44–46
  confidence intervals, 49–52, *51*
  dashboarding frameworks, 30–31
  data visualization, 31, **31–32**
  highlighting seasonality, 42–43
  interactive exploration, 29
  `jointplot` function, 52–55, *54*
  libraries for, **30**
  long-term growth, 42–43
  with Matplotlib, 32–37
  plotting monthly revenue, 42–43
  smoothing trends, 47–49
  static charts, 29
visualizing networks
  financial fraud detection, 250
  force-directed graph, 249
  hubs and bridges, 249
  marketing attribution, 251
  organizational behavior, 251
  portfolio management, 251

weighted graphs, 252–254, *253*